CONTEMPORARY ISSUES in SCIENCE

BIODIVERSITY

CONTEMPORARY ISSUES in SCIENCE

BIODIVERSITY

MIRIAM BOLEYN-FITZGERALD

Biodiversity

Copyright © 2012 by Miriam Boleyn-Fitzgerald

All rights reserved. No part of this book may be reproduced or utilized in any form or by any means, electronic or mechanical, including photocopying, recording, or by any information storage or retrieval systems, without permission in writing from the publisher. For information contact:

Facts On File, Inc.
An imprint of Infobase Learning
132 West 31st Street
New York NY 10001

Library of Congress Cataloging-in-Publication Data
Boleyn-Fitzgerald, Miriam.
  Biodiversity / author, Miriam Boleyn-Fitzgerald.
    p. cm.—(Contemporary issues in science)
  Includes bibliographical references and index.
  ISBN 978-0-8160-6207-2
  1. Biodiversity—Juvenile literature. I. Title.
    QH541.15.B56B67 2011
    333.95'16—dc22   2011001531

Facts On File books are available at special discounts when purchased in bulk quantities for businesses, associations, institutions, or sales promotions. Please call our Special Sales Department in New York at (212) 967-8800 or (800) 322-8755.

You can find Facts On File on the World Wide Web at http://www.infobaselearning.com

Excerpts included herewith have been reprinted by permission of the copyright holders; the author has made every effort to contact copyright holders. The publishers will be glad to rectify, in future editions, any errors or omissions brought to their notice.

Text design by Annie O'Donnell
Composition by Keith Trego
Illustrations by Sholto Ainslie
Photo research by Suzanne M. Tibor
Cover printed by Yurchak Printing, Inc., Landisville, Pa.
Book printed and bound by Yurchak Printing, Inc., Landisville, Pa.
Date printed: November 2011
Printed in the United States of America

10 9 8 7 6 5 4 3 2 1

This book is printed on acid-free paper.

CONTENTS

Preface	ix
Acknowledgments	xiii
Introduction	xiv

1. CONSERVING BIODIVERSITY — 1
The Doomsday Vault: An Experiment in Species Survival	1
Causes of Extinction	5
Ethical, Aesthetic, and Economic Reasons to Conserve	12
Patenting Life: The Commercial Ownership of Genes and Organisms	18
Summary	21

2. HABITAT DESTRUCTION AND RESTORATION — 22
The Polar Bear: Beloved Casualty of Global Warming	22
Thirty-four Conservation Hot Spots Host Half the World's Plant Species	28
The Disappearing Amazon	30
Restoring the Rain Forest	39
Summary	40

3. TOXIC CONTAMINATION AND CLEANUP — 42
Rachel Carson's *Silent Spring*	42
Bald Eagles, California Condors, and the DDT Ban	47
The BP Oil Spill Disaster and Its Aftermath	52
Ten Easy Ways to Contribute to Better Water Quality	59
Summary	67

4. SPECIES OVEREXPLOITATION, PROTECTION, AND CAPTIVE BREEDING — 69

Commercial Whaling Endangers Most of the World's Great Whales — 70
Protecting the Last of the Biggest Big Cats — 76
Is Captive Breeding an Answer for Species Depletion? — 77
A Major Setback for the Gray Wolf, Rocky Mountain Keystone Species — 80
Summary — 82

5. INVASIVE SPECIES AND THEIR IMPACT ON DIVERSITY — 85

Small Snake, Big Fish: Invasive Animals Wipe Out Native Species — 86
Competition from Exotic Species: Friend or Foe to Diversity? — 90
Cutting Down Trees to Restore a Forest — 92
Vacuuming the Reef — 95
Climate Change Encourages the Spread of Invasive Species — 97
Summary — 99

6. ECOSYSTEM DISRUPTION AND THE CALL FOR MEGA-RESERVES — 101

Why Coral Reefs Need Sharks — 102
Elephants, Ants, and Acacia Trees: The Importance of Being Eaten — 106
Demolition of Elephant Culture and Its Brutal Consequences — 108
Mega-reserves: The Best Fix for Ecosystem Disruption — 110
Summary — 113

Chronology	115
Glossary	122
Further Resources	127
Index	145

PREFACE

"Whenever the people are well-informed, they can be trusted with their own government. Whenever things get so far wrong as to attract their notice, they may be relied on to set them to rights."

—Thomas Jefferson

In today's high-speed, high-pressure world, keeping up with the latest scientific and technological discoveries can seem an overwhelming, even impossible task. Each new day brings a fresh batch of information about how the world works; how human bodies and minds work; how human civilization can "work" the world by applying its collective knowledge. Switch on a television news program or the Internet at this very moment—pick up any newspaper or current interest magazine—and stories about health and the environment, worries about national security and violent crime, or advertisements for the latest communication and entertainment gadgets will abound.

Given the nonstop flow of information and commercial pressures, it may seem that a surface understanding of scientific and technological issues is the only realistic goal. The Contemporary Issues in Science set is designed to dispel the myth that a deeper understanding of new findings in science and technology—and therefore considerable power to influence their use—is out of reach of nonspecialists and should be "left to the experts." The set reviews current topics of universal relevance and explores—through the lens of real people's stories—how recent discoveries have changed daily life and are likely to alter it in the future.

Stories featured in the set have received attention in the popular press—often provoking heated controversy at a local, national, and sometimes international level—because beneath the headlines lie sticky questions about how new knowledge should, or should not, be applied, as illustrated by the following examples:

- *Genetic engineering.* The pace of discovery about the human genome and the genomes of other animal and plant species has been breathless since the year 1953, when James Watson and Francis Crick first described the double helix structure of deoxyribonucleic acid (DNA), the chemical substance that acts as a blueprint for building, running, and maintaining all living organisms. In April 2003—a mere 50 years later—sequencing of the human genome was complete. This impressive surge in knowledge about our genes has been accompanied by intense hopes—and intense fears—about newfound technical powers to manipulate the production of life. The tragic death of 18-year-old Jesse Gelsinger in a 1999 gene therapy trial begged obvious questions: Can medical investigators ever obtain truly informed consent from a volunteer when the risks of an experimental procedure are largely unknown? Are the potential benefits of gene therapy worth the unknown public-health risks of altering the human genome using viral vectors? What are the environmental risks of creating transgenic plant and animal species?
- *End-of-life care.* Bold medical innovations like mechanical ventilation, organ transplantation, and tube feeding have saved and improved the lives of millions of patients since the 1950s. A state of profound unconsciousness known as "irreversible coma" first occurred with the ventilator; before its availability, patients without working respiratory systems died from lack of oxygen. Now the bodies of severely brain-damaged and brain-dead people can be maintained indefinitely with a steady supply of oxygen to their living tissues. Theresa Schiavo's case—and other controversial end-of-life cases—shows how loved ones and medical pro-

fessionals try to grapple with agonizing questions like: When are medical interventions extending meaningful life, and when are they inappropriately prolonging death? If a patient's wishes cannot be known with certainty, who should decide her fate?

- *Diversity of life.* Species loss due to human influence has gathered speed in recent centuries, with the current toll at an unthinkable four species an hour. Many conservation biologists argue that we are in the midst of a sixth major mass extinction—one due almost exclusively to human activities. Habitat loss and disruption, global climate change, overhunting, toxic pollution, and the spread of invasive species are among the greatest threats to life on Earth. How can human civilization reduce its footprint on the planet, restore habitats that have been lost or damaged, and reconnect fragments of larger ecosystems? How can beloved and vulnerable species such as the polar bear, Bengal tiger, and blue whale be saved from annihilation? And how can future catastrophic events such as the BP oil spill be prevented?

- *Water.* With "peak water" (the maximum amount of clean, usable water available globally) predicted to occur sometime in the next 25 years, this vital natural resource is certain to be the source of national and regional conflicts. Water plays an essential role not only in living processes but in industrial-scale heating and cooling and in new alternative energy technologies such as coal gasification, hydrogen production, and biofuels conversion. Water also figures highly in global climate change, acting both as a greenhouse gas and as a dynamic heat reservoir. For humankind's clean water requirements, is technological advancement the problem or is it the solution? Will gigantic energy-efficient desalination plants turn countries with ocean coastlines into the new "wet" OPEC, with "clear gold" (water) replacing "black gold" (petroleum) as the preeminent wealth-generating natural resource? Can technological innovation lessen the terrible toll that floods and droughts take on property and human lives?

Whether readers are students considering a career in a scientific or technical field, science or social studies teachers or librarians, or inquisitive people of any age with personal, professional, or political interests in how new knowledge is applied, the Contemporary Issues in Science set places fresh research findings in the context of real-life stories, clarifying the technical and ethical subtleties behind the headlines and supporting an engaged, informed citizenry.

ACKNOWLEDGMENTS

I would like to extend special thanks to the creative team behind these books: executive editor Frank Darmstadt, for his keen editorial eye; literary agent Jodie Rhodes, for making the match; photo researcher Suzie Tibor, for her talent and tenacity in hunting down pictures; Sholto Ainslie for his outstanding work on illustrations; Annie O'Donnell, for designing a visually engaging text; and Peter Faguy, for his shared vision for the project.

I owe a profound debt of thanks to my parents and to the teachers and professional mentors who encouraged me to clear my own path. It is with boundless gratitude and appreciation that I dedicate the set to my husband, Patrick, and sons, Aidan and Finn, for giving me so many good reasons to get out of bed, sit down and write, and love them.

INTRODUCTION

On the morning of Wednesday, April 21, 2010, the United States awoke to the frightening news that late the night before, a violent explosion had rocked an oil-drilling rig off the coast of Louisiana. Eleven workers were missing, and an undetermined amount of crude oil had spilled into the Gulf of Mexico, threatening its rich diversity of wildlife. Still ablaze by Thursday morning, the *Deepwater Horizon* collapsed and sank into the waters of the Gulf, leaving in its wake a five-mile (8-km) oil slick and many unanswered questions.

Who would take responsibility for the accident and its aftermath—BP, the owner of the well, or Transocean, the owner and operator of the drilling rig? Halliburton, the contractor responsible for the cement that failed to contain the explosion, or the federal government, answerable to its citizens for any failures in oversight? How much oil was still flooding from the seabed, and what would be its ultimate environmental toll?

By the time the rig sank, no one could state with any certainty whether the well was still gushing oil; Rear Admiral Mary E. Landry, commander of the Coast Guard response, said that although the oil on the surface appeared to be residual from the explosions, "We don't know what's going on below the surface of the water."

With each passing hour, survival looked less likely for the 11 missing workers. Questions about their ultimate fate were resolved in tragedy when the Coast Guard called a press conference Friday evening. "We have just made a very difficult decision," Rear Admiral Landry said. "After a three-day search covering 5,300 miles, we have reached a point where reasonable expectations of survival have passed."

The immediate toll in human casualties was terribly clear, but the scope of the environmental disaster was much less certain.

The day after the search for the workers ended, robotic underwater monitoring devices discovered oil still surging from the seabed, fouling the wildlife-rich waters of the Gulf. It would be several months and many inaccurate estimates later before the vast extent of contamination could be stated with any confidence by government scientists: Almost 5 million barrels, or 185 million gallons (more than 700 million L) of crude oil had flooded into the Gulf before the well was successfully plugged nearly three months after the explosion, making the *Deepwater Horizon* disaster the largest accidental marine oil spill in history.

The full extent of harm to Gulf wildlife remains unknown many months later, and it may take years before scientists can accurately assess the accident's long-term consequences. Wildlife biologists are certain, however, that petroleum from the spill is likely to persist in the Gulf ecosystem for years to come. They know this because researchers studying other massive spills—the *Exxon Valdez* off the coast of Alaska (1989), the Ixtoc 1 in Mexico (1979), the *Amoco Cadiz* in France (1978), and the *Bouchard 65* off Cape Cod (1974) to name a few—have uncovered evidence of persistent environmental harm decades later.

The BP spill and other stories in this book have received attention in the popular media and provoked controversy at a local, national, sometimes international level. Yet at the heart of each of these stories, underneath any political rhetoric and media hype, lie serious environmental challenges that will influence the survival of countless species, including our own. Students considering careers in conservation biology, ecology, oceanography, and related sciences, as well as environmental law, ethics, and policy, can immerse themselves in these stories to better grasp the scientific and ethical choices involved and the importance of weighing and balancing potential consequences of those choices carefully, on a case-by-case basis.

Biodiversity tells these stories to get at tough questions such as: How can we reduce the human footprint on the planet, restore habitats that have been lost or damaged, and reconnect fragments of larger ecosystems? How can we protect beloved and vulnerable species like the polar bear, the Bengal tiger, and the blue whale from extinction due to grave and immediate threats from overhunting and global climate change? How can we avert future catastrophic (and preventable) toxic events like the BP spill, and how can we avoid the unintended (and often unknowable) consequences of disrupting complex relationships within ecosystems?

These practical and difficult questions link the stories in this volume and force us to confront our scientific, ethical, and spiritual intuitions about what kinds of life deserve consideration and care. The fundamental question underlying the others: Should our moral scope be limited to humans or should it include animals and plants, even the entire natural world? The answer may seem obvious to each of us as individuals, but personal intuitions vary widely and often come into direct conflict. We see these conflicts play out in heated political debates over how to prioritize conservation issues among a vast array of other policy concerns like health care, tax reform, and national defense.

Biodiversity follows the sharp decline in diversity within land and marine ecosystems since the mid-20th century, looks at past and present challenges to maintaining the health of critical ecosystems, and highlights the experimental solutions that are proving most effective at stemming the tide of loss. Chapter 1 tells the story of the Svalbard Global Seed Vault and the worldwide crop crisis it is designed to avert, outlines the five principal human-made causes of extinctions, and traces the historical development of ethical frameworks for different conservation approaches. The remaining chapters tackle the five major causes of biodiversity loss one by one. Chapter 2 considers habitat loss, fragmentation, and disruption; chapter 3 looks at the unrelenting issue of toxic contamination in an industrialized world; and chapter 4 treats the urgent need to protect endangered species from overexploitation. Chapter 5 investigates attempts to curb the harmful effects of invasive species, while chapter 6 brings many of the book's themes together with a look at megareserves as a possible umbrella remedy for many of the causes of ecosystem disruption.

When theoretical debates about ideal environmental outcomes meet the real challenges of caring for an ailing planet, frustration at the slowness of change can lead to a paralyzing sense of apathy. Many citizens simply want to know: What sacrifices are truly required from individuals and from nations in order to save the planet's remaining stores of biodiversity, and how are our skills, time, and energy best spent toward achieving this goal? New generations of researchers, environmental ethicists, policy makers, and nature lovers will need to turn to individual cases like the ones in these pages to identify key issues for dialogue and to establish common ground for action.

Thanks to extraordinarily dire human-made threats like global climate change, habitat loss, and toxic contamination, time is critically short and the call to action urgent—perhaps more urgent than for any other set of scientific and technical challenges facing our species today. Traditional human-centered conservation approaches have failed to shelter much of the natural world from misuse and are proving no match for the tremendous ecological challenges ahead. This volume considers a range of more inclusive, creative, and ambitious solutions, better able to protect the planet's fragile reserves of life because they guard the health of entire ecosystems in all their complexity, rather than attending piecemeal to their individual parts.

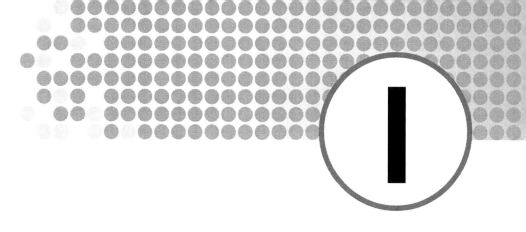

Conserving Biodiversity

This chapter begins with the story of the "doomsday" seed vault—a dramatic attempt on the part of several world governments and private foundations to protect and preserve *biodiversity* in food crops from around the world. Next comes a look at the human causes of shrinking diversity in plant and animal *species* and—at the worst extreme—species *extinctions,* and at the many ethical, economic, and aesthetic arguments for stemming the tide of ecological loss and preserving diversity by protecting individual species and *ecosystems*.

THE DOOMSDAY VAULT: AN EXPERIMENT IN SPECIES SURVIVAL

In February 2008, the Svalbard globale frøhvelv (Global Seed Vault) opened its doors to thousands of seed varieties from all over the world. Nicknamed the "doomsday vault," this futuristic structure was built and funded by the government of Norway as part of a worldwide effort to protect plant species—particularly food crops—from annihilation due to *climate change,* natural disasters, and war. Nestled deep in the heart of a frozen mountain on the remote Norwegian island of Spitsbergen of the *archipelago* of Svalbard—a chilly 800 miles (1,287 km) from the

Svalbard Global Seed Vault (Svalbard globale frøhvelv) in Norway
(© Arcticphoto/Alamy)

North Pole—the vault shelters more than 500,000 unique seed samples at a temperature of -18°C (-0.4°F). The fortlike structure is designed to withstand bomb blasts and earthquakes. It received a major new influx of 50,000 seed samples in December 2009 as world leaders met in Copenhagen, Denmark, to address the climate change crisis.

"We started thinking about this post-9/11 and on the heels of Hurricane Katrina," Cary Fowler, president of the Global Crop Diversity Trust, told the *New York Times* when the structure opened. The Trust, a nonprofit group that operates the vault, is a member of a larger global network seeking to systematically preserve and protect seeds and to gather genetic information from food crops that cannot be stored as seeds—like bananas and coconuts. "Everyone was saying, why didn't anyone prepare for a hurricane before? We knew it was going to happen," Dr. Fowler said. "Well, we are losing biodiversity every day—it's a kind of drip, drip, drip. It's also inevitable. We need to do something about it."

According to the United Nations Food and Agriculture Organization (FAO), there has already been a 75 percent loss in crop diversity over the last century, leaving the food supply exceedingly vulnerable to ecological, political, and climatic changes. In the United States, for example, it is estimated that 7,100 varieties of apples were cultivated in the 1800s; now more than 6,800 of those varieties are extinct. Eighty percent of maize types and 94 percent of pea varieties that existed in the early 1900s are no longer grown. Many modern farms grow just one or two high-efficiency

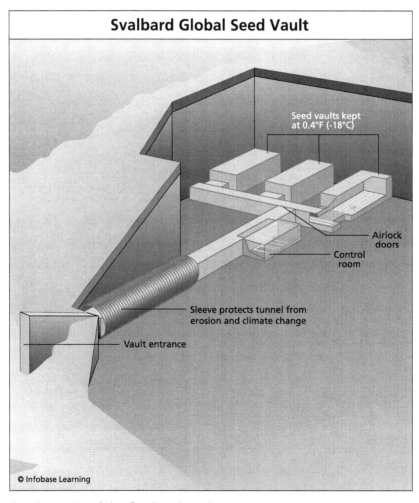

A schematic of the Svalbard vault *(Source: Global Crop Diversity Trust)*

crops, making for increased production—but there is a trade-off. Highly specialized *niche* crops are often the most vulnerable to pests and to even the slightest change in temperature or moisture.

"You need a system to conserve the variety so it doesn't go extinct," Dr. Fowler told the *Times*. "A farmer may make a bowl of porridge with the last seeds of a strain that is of no use to him, and then it's gone. And potentially those are exactly the genes we will need a decade later."

Why house the seeds on an island in the Arctic Circle? It is not especially easy to transport them there; Svalbard's airport is the northernmost airport in the world to receive scheduled flights. On a typical day, a single flight arrives from Oslo on the Norwegian mainland—well over 1,000 miles to the south. The archipelago's capital village of Longyearbyen is also one of the northernmost villages in the world to be inhabited by humans year-round. Home to an *indigenous* population of polar bears, the islands are swallowed in total darkness for almost four months of the year. Despite its long winter night, Longyearbyen sports a human population of just over 2,000.

The icy island location was chosen because it fulfills two major criteria for successful preservation of crop diversity—ideal climate for seed storage and ironclad security. "We are inside a mountain in the Arctic because we wanted a really, really safe place that operates by itself," Dr. Fowler said. Underground in the permafrost near Longyearbyen, the seeds will stay frozen despite any power failures due to natural events, technical failures, or political unrest. The Global Seed Vault is not the first seed bank to exist, but many such facilities have been located in economically unstable or war-torn areas of the world. Seed banks in Iraq and Afghanistan, for example, were raided by looters—not for the seeds themselves, but for the plastic containers they were stored in. Nevertheless, the seed reserves were destroyed in the process. The Global Seed Vault has been described as an insurance policy for these local storage facilities—a centralized backup facility for government and private storage banks all over the world.

The Global Seed Vault was built entirely by the Norwegian government at a cost of 48 million Norwegian kroner (U.S. $9

Emmer wheat, used for ancient Egyptian bread-making, is an example of a threatened crop seed that will be stored in the vault. According to the FAO, three-quarters of the world's crop diversity has been lost in the last century alone. *(Kenneth Garrett/National Geographic/Getty Images)*

million), and its operations and management are financed by the Global Crop Diversity Trust and the government of Norway. The Trust's list of funders includes several national governments (including the United States), international bodies (such as the World Bank and United Nations Foundation), private foundations (including the Bill and Melinda Gates Foundation), and private corporations (such as DuPont).

CAUSES OF EXTINCTION

Over the course of Earth's history, natural events have caused innumerable extinctions, including at least five major *mass extinction events* in the past 540 million years. The largest known mass extinction, the Permian-Triassic event (also known as the "Great Dying"), occurred approximately 250 million years ago when an estimated 95 percent of all marine species and 70 percent of all land species died out. Perhaps the most famous mass extinction (the K-T extinction event) struck 65 million years ago at the end of the Cretaceous period, when an estimated 75 percent of all species became extinct and the 160-million-year reign of the dinosaurs came to an abrupt end.

Many conservation biologists argue that we are currently in the midst of a sixth major mass extinction—one due almost exclusively to human activities. Their claim is persuasive, as species loss due to human actions has accelerated at a shocking rate in recent centuries. It is estimated that, between the years

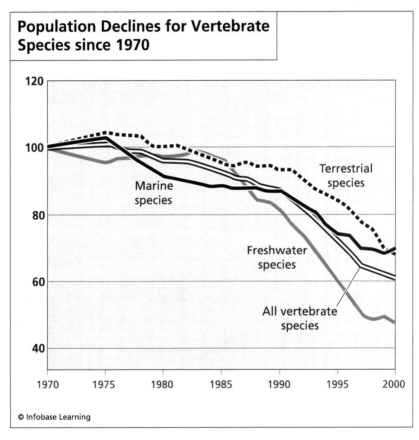

Declines in terrestrial, marine, and freshwater vertebrates over time (as a percentage of their 1970 populations) *(Source: World Wide Fund for Nature and United Nations Environment Programme [UNEP] World Conservation Monitoring Centre)*

of 1600 and 1700, an average of one species of bird and one species of mammal were lost per decade; this average rose to one bird and one mammal per year between 1850 and 1950. Zeroing in on just the four years between 1986 and 1990, this average jumped to four species of each lost every year, and current estimates place the worldwide extinction rate due to human causes at 1,000 to 10,000 times the natural rate.

Some ecosystems are especially rich in plant and animal diversity—the Amazon *rain forest* is the most famous example—and these areas are of special interest to conservation biologists. *Hot spots* like the Amazon are geographic areas of special concern due to a significant level of biodiversity that is threatened with destruction or disruption. Thirty-four hot spots around the world contain approximately half of all *terrestrial* species.

Also of special interest to conservation biologists are particular species that belong to at least one of three categories—*keystone, umbrella,* and *indicator species*—since protection of these animals and plants benefits other species or even entire ecosystems. Keystone species are animal or plant species whose importance to the health of their ecosystems is much greater than would be expected based on a simple analysis of their relative numbers or total *biomass*. (The analogy here is to the keystone concept in architecture—the stone at the crown of an arch that locks the other stones into place. Though the keystone is not necessarily larger or weightier than other pieces, if it is removed, the arch is likely to collapse.)

Umbrella species are species that are selected for special conservation efforts because their land or resource requirements are large enough that protecting them will automatically protect other species living in their *habitat,* while indicator species are animals or plants whose presence, relative abundance, or biological composition are reliable measures of particular environmental conditions. Indicator species are often organisms that are especially sensitive to particular environmental changes and can serve as early warning signals to alert biologists to potential environmental problems such as climate change, disease, or pollution.

8 BIODIVERSITY

The *baiji*, or Yangtze River dolphin, was declared functionally extinct in 2006. The *baiji*'s demise is attributed to an intensive period of hunting by humans in the mid-20th century, industrialization, entanglement in fishing gear, electric fishing (illegal but widely practiced in China), boat collisions, habitat loss, and pollution. *(Kyodo/Landov)*

The five major human causes of extinction are habitat loss, pollution, *overexploitation* of indigenous species, introduction of *invasive species,* and ecosystem disruption. Sometimes when extinctions occur, more than one major factor is at play. Here is a closer look at these five major human sources of ecological stress:

1. *Habitat loss.* When people think of habitat loss, they commonly think of habitat destruction—clear-cut forests or

melting polar ice caps. But habitat loss can also be due to fragmentation (the interruption of larger stretches of habitat with new roads, for example) and disruption. Disruption can be defined as any disturbance in the physical environment in which an animal or plant species lives and can include a wide range of human activities—from dam building, to recreational vehicle use in wild areas, to the many and varied individual and collective behaviors that have led to the recent climate change crisis. (See chapter 2 for more on dwindling natural habitats and attempts to protect them.)

2. *Pollution.* The destruction of multiple bird species due to rampant use of the toxic pesticide DDT spurred the modern environmental movement, thanks to the outspoken environmentalist Rachel Carson and her

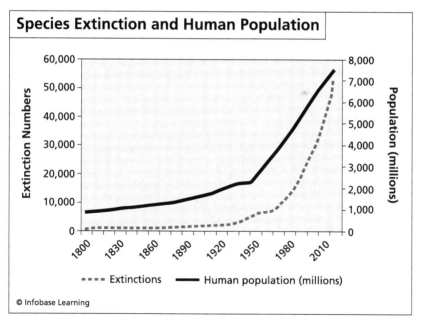

Species extinctions compared with human population growth since 1800, showing a clear relationship between the expanding human footprint and the dwindling diversity of life *(Source: U.S. Geological Survey)*

groundbreaking book *Silent Spring*. How do environmental scientists test for toxic contamination, and how do they go about addressing the problem once it is identified? How can we avert future catastrophic (and preventable) events like the infamous BP spill in the Gulf of Mexico? (See chapter 3 for more on toxic pollution, its detrimental effects, and environmental cleanup.)

3. *Overexploitation.* Species loss commonly occurs when a particular plant or animal population is harvested or hunted by humans to such a degree that their overall population is threatened or extinguished. Famous examples include the overhunting of the North American buffalo and, more recently, the gray wolf; the decimation of great whale populations by commercial whaling; the hunting of elephants for ivory; and the *poaching* of tigers for their bones (for traditional Chinese medicine) and their beautiful pelts. Taking this last example alone, the past century has seen the extinction of three out of nine tiger subspecies—yet illegal poaching remains one of the greatest threats to the tiger's long-term survival, along with fragmentation of its habitat and declining populations of its natural prey. (See chapter 4 for a closer look at the problem of overexploitation and its possible solutions.)

4. *Introduced species.* Recent high-profile examples of the havoc invasive species can wreak on local ecosystems include the brown tree snake, which was introduced to Guam from Australia and was responsible for the *extirpation* of several native species of birds, lizards, and small mammals, and the exploding population of several large and ravenous species of Asian carp in the Mississippi River and its tributaries, which has caused widespread fears that other species throughout the Great Lakes ecosystem might be wiped out. (See chapter 5 for more on invasive species, their impacts on diversity, and approaches to reducing their destructive effects.)

5. *Ecosystem disruption.* What happens to an ecosystem when keystone species are decimated due to habitat

loss, overexploitation, or toxic pollution? Other species dependent on those endangered species suffer, and the delicate food web can collapse altogether. Coral reefs are placed in increasing peril as shark populations dwindle, for example, and the acacia tree—such a vital player in the vulnerable ecosystem of the African savanna—is threatened by the removal of elephants and other large herbivores. (See chapter 6 for an in-depth look at the problem of ecosystem disruption and the protective strategies—including the creation of *mega-reserves*—that are proving most successful.)

Ankole cattle, indigenous to East Africa and bred over the centuries to thrive in the region's harsh conditions, could disappear within 50 years as they are replaced by the imported and more commercially productive Holstein. *(Tasha Lavigne, 2010, used under license from Shutterstock, Inc.)*

ETHICAL, AESTHETIC, AND ECONOMIC REASONS TO CONSERVE

Conservationists and environmental ethicists offer a host of reasons for preserving biodiversity, which can range from *anthropocentric* arguments that elevate human survival, economic well-being, and aesthetic pleasure to the position of greatest importance, to *deep ecology* arguments that shift the focus to the entire natural world, asserting that even nonliving natural things are worth protecting for their own sake. Arguments for preserving biodiversity generally fall into three broad categories along a continuum between the two extremes: anthropocentric, animal rights, and *biocentric*/deep ecology. Here is a closer look at these diverse ethical frameworks.

Protecting Ourselves and Future Generations

Those with strongly anthropocentric views about human civilization's relationship to nature consider humans to be the most important among living beings and therefore subscribe to the belief that our primary responsibility is to each other and to future human generations rather than to other creatures or elements of the natural world. This ethical viewpoint has dominated Western scholarship about the human/environment relationship for centuries, and within an anthropocentric framework, nonhuman beings and natural materials are valuable only as a means to the end of promoting human interests. Aristotle wrote in his foundational work, *Politics,* that "we must suppose . . . that plants are for the sake of animals, and that other animals are for the sake of human beings. . . . If then nature makes nothing incomplete or pointless, it must have made all of them for the sake of human beings."

Many conservationists subscribe to a weaker form of anthropocentrism, one that asserts that it is only possible for humans to understand and interpret the external world from a human point of view. Several influential calls for conservation of natural resources have placed human survival and well-being at the center of ethical consideration, including Gifford Pinchot's conservation ethic. Pinchot was the first chief of the U.S. Forest Service, and he advocated more efficient management and better

A grizzly bear scavenging at a dump station in south-central Alaska. Wild animals foraging through human garbage is one unfortunate consequence of dwindling, fragmented, and disrupted habitats and the resulting decline in their natural food sources. (© Accent Alaska.com/Alamy)

preservation of forests through planned cutting and planting, which he characterized in his 1914 book *The Training of a Forester* as "the art of producing from the forest whatever it can yield for the service of man."

Sometimes the call for the protection of wilderness areas, resources, and biodiversity focuses mainly on direct and indirect economic benefits to humans (as it did with Pinchot and his political champion, President Theodore Roosevelt), and sometimes the rhetorical focus shifts to the aesthetic, health, or spiritual value to humans of preserving the beauty and harmony of nature and humanity's ability to experience and appreciate it.

Protecting Other Sentient Organisms

Most people agree that creating unnecessary suffering for any *sentient* creature is something to be avoided. But people often disagree about what counts as suffering, what creatures (other than humans) feel it, and how the suffering of humans and other animals should be compared in order to decide how much, and what kind, of suffering is necessary.

Andrew Rowan, executive vice president of the U.S. Humane Society, compares the concept of suffering to another slippery, hard-to-describe concept: "Suffering is usually not defined but the usual implication is that it requires a minimum level of cognitive ability that may not be present in most invertebrates (the octopus being a possible exception). The concept appears to be like obscenity where everybody thinks they can recognize it but nobody can define it for regulatory purposes."

Most people believe that animals have some rights, contends Rowan, but most people also believe that they have the right to kill and eat animals. It is a commonly held belief in American culture, grounded in Judeo-Christian teachings, that humans enjoy a divinely granted supremacy over animals and can do to them as they wish, provided their behavior is not cruel or reckless. "Thus," writes Rowan, "whatever 'rights' the public believes animals have claim to, they do not include the right to life." The term *right* in philosophical and legal language is typically used to mean a claim that cannot be trumped simply because it would be useful for someone else to do so. For instance, it is widely believed that humans have the right not to be killed for their organs, even if several lives could be saved by transplants resulting from the taking of that one life.

The strongest animal rights argument, formulated by philosopher Tom Regan in his book *The Case for Animal Rights,* is that nonhuman animals are bearers of moral rights and cannot be used by humans merely as means to the ends of others. Nonhuman animals, Regan argues, are the "subjects-of-a-life"—in other words, just like humans they have a life that matters to them. If we ascribe value to all human beings regardless of their mental capabilities, then we should ascribe it to nonhuman animals as well. A softer animal rights argument is that animals have the right not to be caused suffering.

Australian ethicist Peter Singer, author of the influential 1975 book *Animal Liberation,* contends that "adult apes, monkeys, dogs, cats, rats, and other animals are more aware of what is happening to them, more self-directing, and, so far as we can tell, at least as sensitive to pain as a human infant." His approach falls under the broader ethical category of *utilitarianism*, which seeks to maximize good for the greatest number. The first utilitarian to argue against the infliction of suffering on animals was the 18th-century philosopher Jeremy Bentham, who is widely cited by the animal rights movement. "The question," Bentham wrote, "is not, Can they *reason?* nor, Can they *talk?* but, Can they *suffer?*"

Singer argues that people's willingness to look the other way while animals are harmed and their habitats polluted or destroyed is attributable to *speciesism,* a type of bigotry he likens to racism and sexism, and this sentiment is echoed by many animal rights activists today.

Protecting All Living Things and the Natural World

In a radical shift away from anthropocentric perspectives, biocentric and deep ecology arguments expand the ethical scope to all living things regardless of whether they possess sentience or awareness (biocentrism) or even to the entire natural world, including nonliving substances like water, soil, and rock (deep ecology). Both traditions are rooted in the writings of early naturalists like Henry David Thoreau (1817–62), whose famous *Walden* celebrated the beauty of wild areas and called for a harmonious relationship between humans and the rest of the natural world; John Muir (1838–1914), founder of the Sierra Club and early advocate for the preservation of wild areas through the creation of national parks; and Aldo Leopold (1887–1948), professor of wildlife management at the University of Wisconsin whose influential book, *A Sand County Almanac,* set forth his argument for a *land ethic*—an environmental ethic based on the inherent value of the natural world and its elements rather than on their economic, recreational, or aesthetic value to humans.

"The land ethic," wrote Leopold, "simply enlarges the boundaries of the community to include soils, waters, plants, and animals, or collectively: the land." A simple and straightforward idea, but Leopold knew that adopting a land ethic would require a radical shift in thinking for most people—even many people who believe themselves to be environmentalists. "Do we not already sing our love for and obligation to the land of the free and the home of the brave?" he wrote. "Yes, but just what and whom do we love? Certainly not the soil, which we are sending helter-skelter downriver. Certainly not the waters, which we assume have no function except to turn turbines, float barges, and carry off sewage. Certainly not the plants, of which we exterminate whole communities without batting an eye. Certainly not the animals, of which we have already extirpated many of the largest and most beautiful species."

If our framework for ethical decision-making does not include all elements of the natural world as members of a larger community, Leopold argued, then our relationship to the land will always be one of owner to property, master to slave. "The land-relation is still strictly economic," he wrote, "entailing privileges but not obligations," and he asserted that conservation of natural resources based on their economic value alone is destined to fail because it "tends to ignore, and thus eventually to eliminate, many elements in the land community that lack commercial value, but that are (as far as we know) essential to its healthy functioning." Extending our ethical obligations to the natural world as a whole is the only viable way of adequately protecting it, Leopold believed. "In short, a land ethic changes the role of Homo sapiens from conqueror of the land-community to plain member and citizen of it. It implies respect for its fellow-members, and also respect for the community as such."

(continues on page 20)

(opposite page) Threatened and endangered species of mammals, amphibians, reef-building corals, and cycads, as evaluated by the International Union for Conservation of Nature (IUCN) Red List of Threatened Species. According to IUCN's assessment of the 2008 list, nearly one-quarter of the world's mammals, nearly one-third of amphibians, and more than one out of every eight bird species are at risk of extinction. *(Source: 2008 IUCN Red List)*

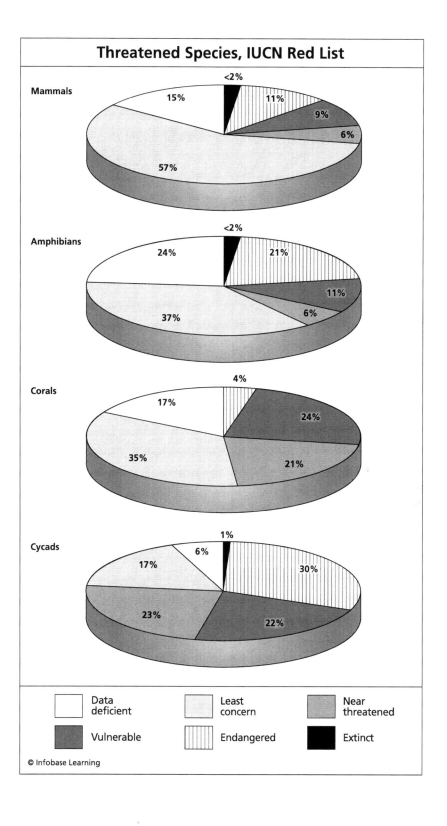

PATENTING LIFE: THE COMMERCIAL OWNERSHIP OF GENES AND ORGANISMS

Until the last few decades, U.S. legal doctrine prohibited patenting any products of nature, but in 1980 the U.S. Supreme Court broke with this tradition by ruling that a genetically engineered bacterium capable of breaking down crude oil was patentable. Chief Justice Warren Burger wrote that "the fact that microorganisms are alive is without legal significance" and that patent law covers "anything under the sun that is made by man." Because the organism had been genetically modified, argued the 5-to-4 majority, it no longer counted as a product of nature.

This trend continued in 1988, when the U.S. Patent Office granted patent number 4,736,866 to the president and fellows of Harvard College for OncoMouse—a genetically modified mouse that was extraordinarily susceptible to cancer. OncoMouse (from *onco,* Greek for "tumor") had been created by researchers at Harvard by introducing a gene that encourages tumor growth into a fertilized mouse embryo. The Patent Office's decision was in keeping with its new policy of taking "non-naturally occurring non-human multicellular living organisms, including animals, to be patentable subject matter."

The controversial practice of patenting living organisms expanded throughout the 1990s to include vast numbers of genes—the blueprints for all living organisms. Biotech companies, universities, and research institutions now own approximately one-fifth of the genes that build and maintain the human body, including genes that can cause obesity, cardiovascular disease, asthma, and breast cancer. Though naturally occurring DNA sequences are certainly considered "products of nature," patents are issued for the chemical sequences discovered in the technical process of decoding DNA. The holder of exclusive patent rights can charge other laboratories licensing fees to use the information in clinical testing.

Michael Crichton, doctor and author of such well-known fiction as *Jurassic Park* and *ER,* noted in the *New York Times* in February 2007 that gene patents "slow the pace of medical advance on deadly diseases. And they raise costs exorbitantly: a test for breast cancer that could be done for $1,000 now costs $3,000. Why? Because the holder of the gene patent can charge whatever he wants, and does. . . . He owns the gene. Nobody else can test for it. In fact, you can't even donate your own breast cancer gene to another scientist without permission. The gene may exist in your body, but it's now private property."

Peter Shorett, director of programs for the Council for Responsible Genetics, calls patents on genes and organisms a "'toll booth' through which future scientists must pass" and says that the higher the cost of obtaining model organisms like OncoMouse, "the more biomedical innovations will be impeded, as researchers in the early stages of their work may choose to look elsewhere, not willing to pay steep up-front costs or abide by unyielding restrictions." The patent process also hinders a "primary mission" of universities, Shorett argues—the free and open exchange of knowledge. "Secrecy and under-communication become the norm as faculty members withhold data from the scientific community to protect proprietary interests."

Academic and industrial secrecy restrict not only biomedical innovation, but also conservation efforts. Bent Skovmand, a scientist who helped create the Svalbard Global Seed Vault, devoted much of his life to tracking down and preserving lost strains of food crops. He gathered more than 150,000 kinds of wheat and more than 20,000 kinds of corn before his death in 2007, and he spoke out against the patenting of individual plant genes, arguing that there should be free and public access to such basic genetic information. He told the *New York Times* in 2000 that copyrighting genes is "like copyrighting each and every word in 'Hamlet,' and saying no one can use any word used in 'Hamlet' without paying the author."

Dr. Skovmand made a point of putting all his own data on CDs and distributing them for free. He said he considered

(continues)

(continued)

labeling them with the following message: "Duplication of this CD is enthusiastically encouraged."

In October 2010, the U.S. Justice Department sided with plaintiffs in a lawsuit that challenged patents on two genes implicated in breast and ovarian cancers, BRCA1 and BRCA2, arguing that the "chemical structure of native human genes is a product of nature, and it is no less a product of nature when that structure is 'isolated' from its natural environment than are cotton fibers that have been separated from cotton seeds or coal that has been extracted from the earth." While this is a major reversal of federal policy, it is unclear how the new position might change the Patent Office's historical practice of issuing patents on genes. Moreover, in the Justice Department brief, the federal government defended the rights of biotechnology companies to continue to patent manipulated DNA, including gene therapies and genetically engineered crops.

(continued from page 16)

In 1973, mountaineer and philosopher Arne Naess introduced the term *deep ecology* to environmental literature, contrasting it with more shallow mainstream ecological approaches that tend to look for short-term fixes and consumption-driven answers to environmental problems, rather than deep questioning that might lead to the fundamental, long-term changes necessary to preserve the ecological diversity of natural systems.

Deep ecology scholar Alan Drengson of the University of Victoria in British Columbia writes that contrary to popular belief, "Supporters of the deep ecology movement are not antihuman, as is sometimes alleged." He points out that the movement's platform statement, drafted by Naess and fellow ecologist George Sessions, starts with the recognition of the inherent worth of all beings, including humans: "The well-being and flourishing of human and nonhuman life on Earth have value in themselves," wrote Naess and Sessions. "These values are independent of the usefulness of the nonhuman world for human purposes."

SUMMARY

Given the stunning rate of species loss—an estimated four species each hour—it is easy to feel discouraged. Even if one day it becomes technically feasible to reproduce lost animal species from preserved DNA samples à la *Jurassic Park*—and even if the Svalbard Global Seed Vault succeeds in its mission to preserve seeds from the majority of existing food crops—the genetic blueprints for most extinct species of plants and animals are already lost forever. Moreover, even if we could repopulate many threatened species, the challenge of restoring vast stretches of lost habitats is huge.

No doubt the planet's enduring biodiversity is under critical and immediate threat, but the situation is far from hopeless. Each major danger to individual species and larger ecosystems has at least one workable answer, and conservation biologists are teaming up with environmental engineers and lawmakers around the globe to compare possible solutions and implement the most promising conservation strategies. Does captive breeding work best to protect an endangered species of fish, for example, or are there creative ways to turn fishermen into conservationists? Is it enough to prohibit hunters from killing tigers for their bones and pelts, or are mega-reserves necessary to ensure tigers' long-term survival? These and other questions are addressed in the compelling case studies that follow.

2

Habitat Destruction and Restoration

The most serious and immediate threat to the planet's rich diversity of plant and animal species—and to indigenous human cultures—is the loss of natural habitats through ecosystem destruction (such as clear-cutting and burning forests, filling in *wetlands,* or damaging coral reefs), fragmentation (such as road-building through large stretches of habitat), and disruption (such as dam-building or off-road vehicle use). As the human footprint on the planet expands at an unprecedented rate, a vast number of flora, fauna, and indigenous cultures face imminent extinction. This chapter looks at two interrelated ecological crises that are top priorities for international conservation efforts: habitat loss due to global climate change and the reckless destruction of tropical rain forests.

THE POLAR BEAR: BELOVED CASUALTY OF GLOBAL WARMING

On August 16, 2008, federal contractors performing an aerial survey in the Chukchi Sea off the northwest coast of Alaska were on the lookout for bowhead whales—a protected species—as part of preparatory work for offshore oil exploration and development. From their vantage point over the water, they

Habitat Destruction and Restoration

found something unexpected: nine polar bears swimming in the open sea miles from the Alaskan shore, one of them a full 65 miles (105 km) from land. At least some of the bears were swimming north, apparently in a futile attempt to reach the closest stretch of polar ice, which—due to a devastating arctic warming trend—was a full 400 miles (644 km) away.

Until the past few years, it was a rare event to see polar bears swimming so far from shore. During the 16-year stretch between 1987 and 2003, routine marine mammal surveys identified only a total of 12 bears swimming in the open sea. Sadly, sightings have become much more common as arctic ice retreats and the summers grow longer. Geoff York with the World Wildlife Fund's (WWF) Arctic Programme commented on the implications of bears swimming so far from land. "To find so many

A polar bear on an ice floe *(Jan Martin Will, 2010, used under license from Shutterstock, Inc.)*

Comparison of Arctic Sea Ice Summer Minimums, 1980 and 2007/2008

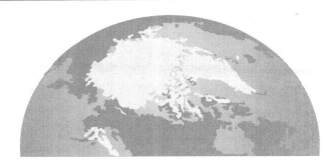

1980 summer minimum

2007 summer minimum

2008 summer minimum

© Infobase Learning

Satellite images show that the Arctic has been losing summer sea ice for the past 30 years, and computer models predict that the loss will continue. *(Source: NOAA)*

polar bears at sea at one time is extremely worrisome because it could be an indication that as the sea ice on which they live and hunt continues to melt, many more bears may be out there facing similar risk," he said. "As climate change continues to dramatically disrupt the Arctic, polar bears and their cubs are being forced to swim longer distances to find food and habitat."

Polar bears are extraordinarily good swimmers, and healthy adult bears have been observed swimming distances up to 100 miles (160 km) to reach the arctic ice, their primary hunting ground. But these marathon swims leave the bears depleted and susceptible to drowning—especially when a storm arises. According to WWF, several polar bear carcasses have been observed floating in open water in recent years, presumably after drowning.

Arctic regions are seeing the most acute and dramatic effects of global climate change, and those effects are accelerating, according to the National Oceanic and Atmospheric Administration (NOAA). NOAA reports that arctic ice is melting faster than predicted by the most recent analysis by the Intergovernmental Panel on Climate Change (IPCC), "Climate Change 2007," the IPCC Fourth Assessment Report, the most recent analysis performed by that international panel of climate scientists. (As of 2011, the IPCC Fifth Assessment Report is under way, with 831 highly qualified experts chosen and working groups scheduled.) "Arctic sea ice extent observed by satellites has been shrinking for the past 30 years," NOAA stated in a 2009 fact sheet on the future of arctic sea ice. Ice coverage in 2007, 2008, and 2009 was the lowest since the satellite record began in 1979, and though the 2009 summer sea ice cover was more than it was in 2007 and 2008, it was still 25 percent less than the average for the period between 1979 and 2000.

As the summer melting season warms and lengthens, there is less sea ice at summer's end, and since 2005 the fall freeze has been delayed significantly, as summertime absorption of solar energy in open water pumps up the overall heat content of the ocean. "Warming is enhanced near the surface and into the lower atmosphere," NOAA reported, "and over 50% of the thick sea ice that was built up over many winters (nearly 10 feet

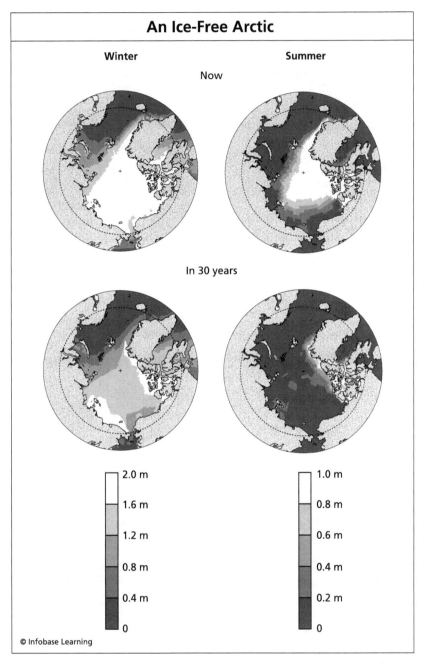

The bottom images show computer models of a nearly ice-free Arctic in summer, projected to occur within 30 years if current warming trends continue. *(Source: M. Wang and J. E. Overland)*

[3 m] thick) has melted and is replaced with thinner ice formed in a single year. The increased absorption of heat from the sun into the ocean and the atmosphere is a major rapid contributor to continued sea ice reduction."

With the 2007 and 2008 summer sea ice levels as a starting point, computer models now predict that the Arctic could be nearly sea ice–free in the summertime within a mere 30 years (see illustration on page 26). This has devastating implications for polar bears, since they cannot find the high-energy diet they require (primarily ringed and bearded seals) by hunting on land. Some scientists predict that if current melting trends continue, three of four major polar bear populations will be extinct by 2075.

In 2005, the Center for Biological Diversity, Greenpeace, and the Natural Resources Defense Council (NRDC) filed a suit demanding that the polar bear be listed under the Endangered Species Act. After years of delay, the Interior Department of President George W. Bush agreed to the listing in May 2008, while also allowing oil and gas exploration and development to continue in the bears' habitat so long as energy companies complied with requirements of the Marine Mammal Protection Act. The administration postponed establishing a *critical habitat*—the area deemed necessary for the survival of the species. Delineation of a critical habitat is a requirement of the Endangered Species Act, and a 2009 *New York Times* editorial concluded that the Bush administration delayed the designation "partly because doing so could have torpedoed its grand plans to open millions of acres of prime polar bear territory in the Beaufort and Chukchi Seas to oil and gas exploration."

In October 2009, the Obama administration announced a proposal to designate 200,000 square miles (518,000 km^2) of Alaskan sea and sea ice as critical habitat for the polar bear. The announcement came after the secretary of the interior, Ken Salazar, angered environmentalists by upholding a Bush administration finding that the Endangered Species Act is not an appropriate tool with which to enforce restrictions on CO_2 and other *greenhouse gas* emissions, asserting that other national legislation

(continues on page 30)

THIRTY-FOUR CONSERVATION HOT SPOTS HOST HALF THE WORLD'S PLANT SPECIES

In 1988, British environmentalist Norman Myers identified 10 tropical forest hot spots—areas he singled out for special attention in conservation efforts. These areas of the world should be prioritized for two reasons, Myers argued: First, because they contain an exceptionally rich array of plant diversity, and second, because they have already been radically reduced in size by human activities like logging, ranching, and urban and suburban sprawl.

The new concept was compelling to conservationists but still quite loosely defined when the environmental organization Conservation International (CI) began working with Myers in the late 1990s to establish clear quantitative criteria by which hot spots could be identified and to perform an extensive global review employing the new criteria. To qualify as a hot spot under the more specific guidelines, a region needed to contain at least 1,500 species of *vascular plants* (greater than 0.5 percent of the world's total) as *endemic* species (species unique to that particular region) and its geographic area must have already been slashed by at least 70 percent.

A second analysis using these strict quantitative guidelines was published in the year 2000 in the journal *Nature,* and it documented a total of 25 biodiversity hot spots. Together, these areas contained 44 percent of the world's endemic plant species and 35 percent of terrestrial vertebrate species in an area that formerly covered approximately 12 percent of the planet's land area. However, after being slashed by nearly 90

(continues)

(opposite page) A map of the world's 34 terrestrial hot spots—severely threatened land areas containing an exceptionally rich array of plant diversity

Habitat Destruction and Restoration 29

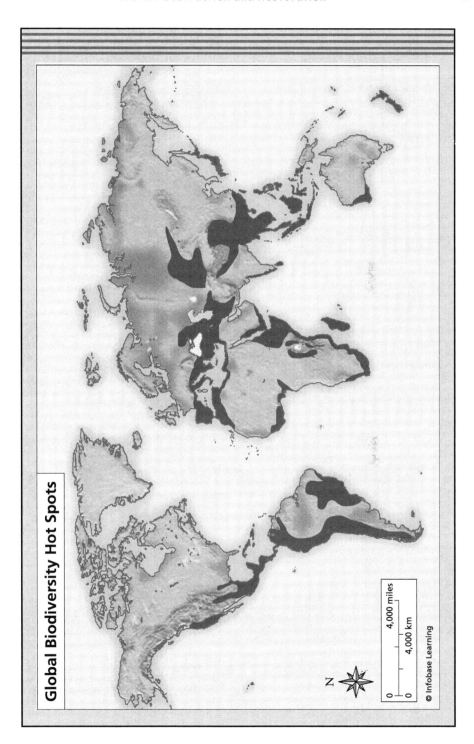

(continued)
percent, this treasure trove of biodiversity was now restricted to a mere 1.4 percent of the planet's land area.

"Conservationists are far from able to assist all species under threat, if only for lack of funding," Myers and colleagues wrote in the 2000 *Nature* analysis. "This places a premium on priorities: how can we support the most species at the least cost?" The best way, they argued in this influential article, would be to focus on these biodiversity hot spots "where exceptional concentrations of endemic species are undergoing exceptional loss of habitat," thus facilitating what they called a "silver bullet" strategy for conservationists, "focusing on these hotspots in proportion to their share of the world's species at risk."

In their 2005 update, *Hotspots Revisited,* CI identified an additional nine critical areas, bringing the total to 34 hot spots that once covered 15.7 percent of the Earth's land surface but covered a mere 2.3 percent in 2005. When that update was performed, these areas contained a staggering percentage of the world's species as endemics: about 50 percent of the world's plant species, 42 percent of terrestrial vertebrate species, and 29 percent of freshwater fish species.

Dr. Agnes Kiss, an environmental specialist with the World Bank, told the *New York Times* in July 2003 that the hot spot concept has made it much easier to obtain funding for con-

(continued from page 27)
(such as the Clean Air Act) and international agreements (such as the Copenhagen Accord discussed later in this chapter) were more suited to that purpose.

THE DISAPPEARING AMAZON
On February 12, 2005, Sister Dorothy Stang, a Catholic nun from Dayton, Ohio, and an outspoken defender of the Amazon rain forest and its native peoples, was shot and killed on a jungle

servation proposals involving these regions of the world. "Put it this way," she said. "When we're trying to justify a project, if it's a hot spot, basically it's a shoo-in."

Some environmentalists, however, argue that the strategy of prioritizing funding for hot spots is not without its own serious risks. Critically important natural areas that should be prioritized for legal protection and funding will be missed by the hot spot model altogether. "The hot-spot concept has grown so popular in recent years within the larger conservation community that it now risks eclipsing all other approaches," wrote Dr. Michelle Marvier, a biology professor at Santa Clara University, and Dr. Peter Kareiva, a scientist with the Nature Conservancy, in a July 2003 paper published in *American Scientist.*

Wetlands are critically important, for example, because they prevent disastrous flooding, form a natural water filtration system, and play host to fish nurseries. They would not, however, qualify for protection as hot spots since they are home to fewer species. The hot spot model is also land-based and therefore does not address biodiversity loss in the oceans.

Russell Mittermeier, president of CI, says hot spots have been successful in attracting attention and financing for conservation in tropical countries. "And that has been good," he said. "No one is suggesting that one invest solely in hot spots, but if you want to avoid extinctions, you have to invest in them."

road in the Brazilian state of Pará after blocking ranchers from taking over a stretch of rain forest that was home to 400 families. Several men were later convicted of the 73-year-old nun's murder, including rancher Vitalmiro Bastos de Moura, who was found guilty of ordering the killing.

The world was shocked by the brutality of the act, but sadly, her fellow activists were not. Sister Stang had received many death threats for her activism on behalf of the forest and its poorer residents, and more than 1,000 settlers, union members,

The Nanay River, a tributary of the Amazon River, winding through the Peruvian rain forest *(© John Warburton-Lee Photography/Alamy)*

and priests had been killed in Pará in land disputes during the preceding 30 years. Tarcísio Feitosa da Silva, director of the Roman Catholic Church's Pastoral Land Commission in Pará, told the *New York Times,* "She aroused the ire of a lot of people by discovering all those irregularities that were going to damage a lot of big interests." When asked if he believed that he himself had angered those same interests, da Silva answered, "Maybe so, but they are going to have to leave. The forest must be kept alive for the benefit of all, not just a few speculators."

In the month before Sister Stang's killing, tensions had been on the rise after Brazil's government had announced new regulations for land use and ownership. Ranchers and loggers had blocked roads and burned buses, and they had warned that "blood will flow" if the government did not loosen the regulations. In response to threats of violence, the Brazilian government had effectively reopened the land to development by delaying enforcement of the new rules.

Habitat Destruction and Restoration

For decades, the Amazon rain forest has inspired unshakable passion in its protectors and violent greed in many of those who seek to exploit it for monetary gain. It is the mother of all biodiversity hot spots, with at least one in 10 of all known species on Earth living within its lush and fertile territory. (The inaccessibility of parts of the forest and the denseness of life make cataloguing difficult, but most estimates place the species count in the ballpark of 10 to 30 percent of the world's total.) Home to a staggering 40,000 catalogued plant and tree species and source of one-fifth of the world's free-flowing freshwater, the Amazon rain forest shelters the densest diversity of birds, butterflies, and freshwater fish anywhere on Earth. The rain forest also boasts an impressive array of cultural diversity, with an estimated 349 ethnic groups living within its boundaries and speaking at least 300 different dialects.

Brazilian farms encroach on the Amazonian rain forest (September 2008) to clear the way for more cattle ranches and soybean farms. *(John Stanmeyer/VII/Corbis)*

The Extent of the Devastation

An estimated 17 to 18 percent of the Amazon rain forest's original cover has already been destroyed, the majority cleared for cattle ranching. In the 2006 report "Livestock's Long Shadow," the Food and Agriculture Organization (FAO) of the United Nations notes that global livestock grazing and food production use 30 percent of the planet's land area and that the expansion of grazing land destroys more rain forest than any other human activity or natural process. Other human behaviors that worsen the threats to this complex and fragile habitat include mining,

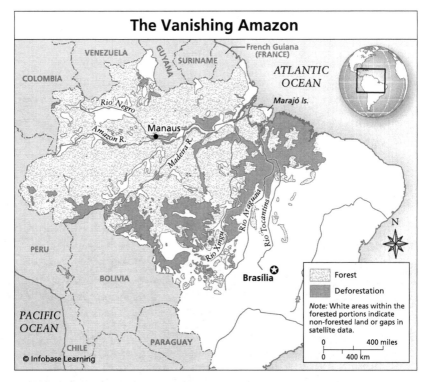

Deforestation in the Brazilian Amazon, as of 2008. The practice of burning trees to clear stretches of rain forest for ranching and farming has emitted enormous amounts of carbon dioxide to date, but new conservation measures enacted by the Brazilian government are slowing the pace of destruction. *(Source: Brazil National Institute for Space Research)*

Habitat Destruction and Restoration

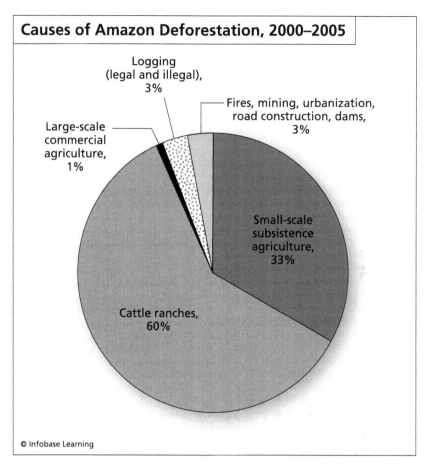

This graph of the causes of deforestation in the Amazon shows that cattle ranching is by far the primary source of destruction. *(Source: Mongabay.com)*

logging, road-building, and burning forests to make charcoal power for industrial plants.

The Amazon rain forest is the largest CO_2 reserve on earth, and with every tree lost the lungs of the planet grow smaller. Thomas Friedman, op-ed columnist for the *New York Times*, related the sobering statistics this way in November 2009: "Imagine if you took all the cars, trucks, planes, trains and ships in the world and added up their exhaust every year," he wrote. "The amount of carbon dioxide, or CO_2, all those cars, trucks, planes, trains and ships collectively emit into the atmosphere is actually less than the carbon emissions every year that result from the chopping down and clearing of tropical forests in places like Brazil, Indonesia and the Congo. We are now losing a tropical forest the size of New York State every year, and the carbon that releases into the atmosphere now accounts for roughly 17 percent of all global emissions contributing to climate change."

Deforestation's contribution to climate change qualifies as a global emergency, and the cost in biodiversity has also been extreme. An August 2008 study in the *Proceedings of the National Academy of Sciences* (*PNAS*) projected that 20 to 33 percent of the canopy tree species in the Amazon will be extinct by 2050 thanks to deforestation, while a December 2008 study (also published in *PNAS*) concluded that about 9 percent of all vascular plants would be gone by 2050. Whichever figure ends up falling closer to the mark, conservation biologists agree that the losses will be severe and irretrievable.

Cashing In on Preservation

One small, rain forest–rich country has a solid plan to turn the tide of destruction. Guyana—Brazil's neighbor to the north along the Atlantic coast—earned itself international recognition at the December 2009 United Nations Conference on Climate Change in Copenhagen for its cutting-edge plan for forest preservation and low-carbon development. Guyana is roughly the size of Great Britain, and its land area is still 75 percent rain forest. Its president, Bharrat Jagdeo, is an economist by training, and he has initiated a pilot effort (largely funded by the Norwegian government) to earn international assistance for rain forest preservation and *sustainable development.*

The plan works like this: A developing country like Guyana calculates the benefit of deforestation to its economy—in other words, the income it would receive for timber and for meat and agricultural products from ranches and farms. (In Guyana's case, McKinsey, the international management consulting firm, calculated this benefit to be $580 million per year for more than 25 years.) Rather than pursuing deforestation as a development

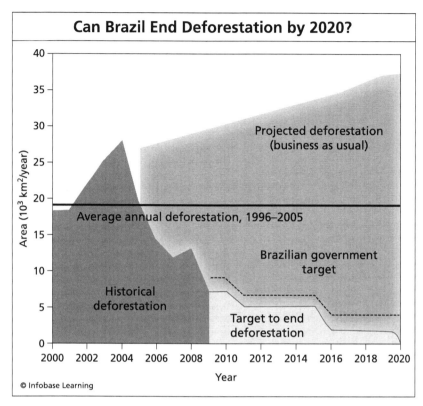

This figure shows historical deforestation rates in the Brazilian Amazon, as well as projections for future deforestation under three possible scenarios: business as usual (assuming a continuation of the economic and regulatory trends as of 2003); the Brazilian government's current target; and a more radical plan that could put an end to deforestation by 2020. *(Source: D. Nepstad, et al., 2009)*

plan, these countries then put forward sustainable development plans that richer nations could invest in—based on the premise that rain forests provide critical environmental services to the world (such as *carbon sequestration,* oxygen generation, rain production and maintenance of stable rainfall patterns, and biodiversity protection) and that, as rain forest ecosystems shrink, these services provided by standing forests will only increase in economic value.

"The forest should be seen like a giant public utility," Andrew Mitchell, director of the Global Canopy Program, told the *New York Times* in December 2009. The Global Canopy Program is a cooperative effort among 37 scientific institutions devoted to forest canopy research, education, and conservation. "The forest-owning nations, instead of receiving aid, would become suppliers of a global service that is paid for or rented by the international community," he told the *Times*. "It changes from an aid relationship into a trade relationship."

At the Copenhagen meeting, Guyana's pilot program was cited in support of a broader international program known as UN-REDD (The United Nations Collaborative Programme on Reducing Emissions from Deforestation and Forest Degradation in Developing Countries), which would allow industrialized nations to meet CO_2 commitments by funding forest preservation projects in other countries. The new treaty, the Copenhagen Accord, endorses the use of the UN-REDD program to reduce deforestation and recognizes "the crucial role of reducing emission from deforestation and forest degradation and the need to enhance removals of greenhouse gas emission by forests." Unlike the earlier, much weaker Kyoto Accord, the agreement requires concrete CO_2 reduction commitments from participating countries, and by early 2010, 55 countries (including the United States) representing nearly 80 percent of all global CO_2 emissions had submitted their pledges.

Carbon-trade relationships would bring much-needed cash to forest-rich nations like Guyana, while lessening the immediate pressure on industrialized nations to reduce their own carbon emissions by allowing them to offset those emissions with investments in forest preservation abroad. A similar cap-and-trade plan is currently under consideration by the U.S. Congress. These innovative approaches reflect the preservation philosophy of an influential environmental activist of the 1980s, Francisco Alves Mendes—better known as Chico Mendes—who argued that a standing rain forest could have greater economic value than a forest cut or burned for short-term economic gain. Mr. Mendes, like Sister Stang, was gunned down by ranchers who were opposed to his efforts to save the Amazon.

RESTORING THE RAIN FOREST

Once a stretch of lush tropical forest is cut and developed, what are the best methods of restoring it to a complex forest ecosystem? This question is driving a long-term experimental reforestation project based in the state of Veracruz on the Mexican Gulf coast. Led by University of Illinois at Chicago (UIC) restoration biologist Henry Howe and funded by the National Science Foundation (NSF), the project is designed to compare the most common approach to reforestation with a different method, one that Howe and his colleagues consider to be much more effective at restoring a rich level of diversity to these critical tracts of land.

The traditional method involves planting a small number of quick-growing trees whose seeds are carried by the wind. This approach tends to create *monocultures*—areas dominated by a single plant species—which in turn only support a few animal species. Animal-dispersed trees such as fruit trees, on the other hand, will attract birds and bats back to the replanted area. These animal visitors will drop a variety of seeds that they have carried from deeper forest areas onto the newly planted area and then they will transport fruit seeds from the new trees into the surrounding landscape.

"In the tropics, things grow fast," Dr. Howe said at the start of the project in 2005, in a press release from UIC. "By the end of the first five years, we'll begin to see the influence of animal and wind dispersal. In 10 years, we'll see a very strong effect of seed fall, germination and establishment. Once these stands of animal-dispersed trees have grown up and start producing fruit, they'll be exporting their seeds to surrounding landscapes."

Howe believes that over the years, evidence will pile up in support of the conclusion that animals like birds and bats transport a diverse array of tree seeds, including many deep-forest varieties, and that this method of reforestation will support a greater degree of plant and animal diversity. "It's a way of providing connection between existing, isolated trees, remnants and fragments and large tracts growing along rivers in rain forests," he said.

The diversity-promoting role of our winged friends can also be seen on shade-grown coffee farms, according to a December 2008 study by University of Michigan researchers Shalene

Jha and Christopher Dick. A university press release about the study noted that shade coffee farms, which grow coffee bushes under a lush canopy of multiple tree species, "not only harbor native birds, bats and other beneficial creatures, but also maintain genetic diversity of native tree species and can act as focal points for tropical forest regeneration."

Many of the farms were clear-cut and burned several decades ago and then replanted with coffee bushes and canopy tree species, including fruit trees. Over time, birds flocked to the plantations and carried seeds from *understory* tree species onto the farmland from "mother trees" in the forest. Farmers then allowed these trees to flourish over the years because they play an important role in preventing soil erosion. The conclusion: that "shade coffee farms, by being hospitable to birds, support widespread dispersal of native trees, in effect connecting patches of surrounding forest."

The demand for coffee is greater than ever and is growing by the day. Jha emphasizes the need to take ecological benefits into account when considering how coffee is produced. "A lot of the rustic coffee farms are turning into Sun-intensive operations, where farmers cut down the overstory and try to level out the fields so it's easier to get machines in," she said. "It's more essential than ever to pay attention to the ecological benefits shade coffee farms provide."

Consumers can slow the alarming trend toward Sun-grown coffee by limiting their intake to beans produced on traditional, shade-grown coffee plantations.

SUMMARY

Most species on the planet have adapted over the course of thousands—even millions—of years to live and breed in very particular habitats and climates. When those conditions vanish, so will the species. Polar bears are a majestic and beloved example of this tragic natural fact. They have evolved to hunt seals through polar ice, and if the ice disappears, a species tough enough to thrive in the Polar North will quickly succumb to drowning and starvation.

Conservation biologists agree that many of the world's most critical ecological niches are reaching a tipping point of destruction, beyond which the loss of habitats will accelerate. Polar ice habitats and the Amazon rain forest are among the highest-profile examples. A February 2010 analysis by the World Bank, for example, found that the Amazon rain forest is fast approaching a critical number of 20 percent territory loss and that the resulting climate changes, altered rainfall patterns, and fires could combine with already devastating pressures to shrink the forest to one-third of its original territory by 2075.

The present precipitous rate of habitat loss ranks as an ecological emergency of staggering proportions, and the need for sustainable, low-carbon alternatives to the planet's current, unsustainable course is urgent. But national and international initiatives such as the UN-REDD program could stem the tide of loss and set the globe on a greener, more sustainable course. A December 2009 report by tropical forest ecologist Daniel Nepstad and colleagues, published in the journal *Science,* mapped a plausible scenario for Brazil under which net rain forest loss in that country could be eliminated by 2020—*if* Brazil cashes in on unprecedented opportunities like UN-REDD. Together with forest restoration efforts, a sharp drop in new rain forest loss would greatly alleviate the climate change crisis, while also safeguarding the world's richest store of land-based biodiversity.

3

Toxic Contamination and Cleanup

One of the most daunting challenges to the planet's health is the problem of toxic pollution: How can we clean up environmental mistakes of the past and prevent future disasters? How can we remove deadly chemicals that have been carelessly released into the environment, and how can we avert future catastrophic—and preventable—events like the infamous BP spill in the Gulf of Mexico? This chapter addresses these tough questions, first through a historical lens with a look at the environmental legacy of Rachel Carson and her crusade against toxic pesticide pollution, and then with a review of the recent BP oil spill, its devastating environmental impact, and its implications for the future health of our oceans and the choices required of us all—individually and collectively—to protect them.

RACHEL CARSON'S *SILENT SPRING*

The modern environmental movement owes many of its successes to Rachel Carson (1907–64), an American naturalist and science writer who brought public attention to the dangers of rampant pesticide use in her influential book, *Silent Spring*. After 15 years as a scientist and editor for the U.S. Fish and Wildlife Service (FWS), Carson left her position as editor in chief at that agency to devote herself full time to nature writing. She chronicled the life of the sea in a critically acclaimed trilogy on marine biology *(Under the Sea-Wind, The Sea Around Us,* and *The Edge of the Sea)*, and then turned her attention to the reckless use of dichlorodiphenyltrichloroethane (DDT) and other toxic pesticides, their harmful effects on many bird species, and their potential threats to human health. She argued that humans had "allowed these chemicals to be used with little or no advance investigation of their effect on soil, water, wildlife, and man himself. Future generations are unlikely to condone our lack of prudent concern for the integrity of the natural world that supports all life."

Former vice president Al Gore, in an introduction to the book's newest edition, celebrated *Silent Spring*'s legacy because it "brought environmental issues to the attention not just of

A plane dusts sheep with 10 percent **DDT** powder on a ranch in Oregon in 1948. *(AP Images)*

A woman sprays her bedroom with the army's DDT bomb to kill insects in Brooklyn, New York, in 1945. *(© Bettmann/CORBIS)*

industry and government; it brought them to the public. . ." Gore further credited Carson with bringing us "back to a fundamental idea lost to an amazing degree in modern civilization: the interconnection of human beings and the environment." Environmental engineer and Carson scholar H. Patricia Hynes wrote in her 1989 tribute to the book that as a call to action, *Silent Spring* "altered the balance of power in the world. No one since would be able to sell pollution as the necessary underside of progress so easily or uncritically."

Carson was deeply skeptical of an increasing reliance on specialists, "each of whom sees his own problem and is unaware of or intolerant of the larger frame into which it fits." The danger of blind reliance on so-called experts, she suggested, lay in potential biases that could arise when the interests of big agriculture and industry are tangled up in what should be purely scientific decisions, and she characterized the time as "an era dominated by industry, in which the right to make a dollar at whatever cost is seldom challenged." She was especially critical of the dual role that the U.S. Department of Agriculture (USDA) played as both promoter of agricultural interests and regulator of the use of agricultural pesticides, which she saw as a clear conflict of interest. In 1963, she testified before Congress, calling for the registration of pesticides and other toxic chemicals and for restrictions on their sale and use.

Following the publication of *Silent Spring* in 1962, a freshly educated public called for a total ban on DDT and for the creation of a new independent environmental oversight agency, and in December 1970, the Environmental Protection Agency (EPA) opened its doors. A history of that agency, published in the November 1985 *EPA Journal,* held that "*Silent Spring* played in the history of environmentalism roughly the same role that *Uncle Tom's Cabin* played in the abolitionist movement. In fact, EPA today may be said without exaggeration to be the extended shadow of Rachel Carson."

But Carson's ideas were not universally embraced. She was widely criticized by industry experts as biased, hysterical, and unscientific, and in 1999 naturalist and activist Peter Matthiessen chronicled in *Time* magazine that even before *Silent Spring* was published, there was strong opposition to it. "Carson was violently assailed by threats of lawsuits and derision," he wrote, "including suggestions that this meticulous scientist was a 'hysterical woman' unqualified to write such a book. A huge counterattack was organized and led by Monsanto Company, Velsicol, American Cyanamid—indeed, the whole chemical industry—duly supported by the Agriculture Department as well as the more cautious in the media." A former chemical industry spokesman, Robert White-Stevens, went so far as to claim that "If man were to follow the teachings of Miss Carson, we would return to the Dark Ages, and the insects and diseases and vermin would once again inherit the earth."

Rachel Carson at the seaside in 1961 *(Alfred Eisenstaedt/Time Life Pictures/Getty Images)*

DDT was banned in 1972, and those catastrophic predictions have not come to pass; on the other hand, DDT's long-lasting effects on species at the top of the food chain are still being discovered today. As recently as November 2010, residual DDT in the environment was identified as the most likely reason that populations of the California condor—the nation's largest bird and a longtime resident of the Endangered Species List—are laying eggs with shells so thin that chicks simply cannot survive.

BALD EAGLES, CALIFORNIA CONDORS, AND THE DDT BAN

DDT is an organic compound (a compound built on a foundation of carbon molecules) and can be stored in high concentrations in the bodies of humans and other animals. The unique carbon- and chlorine-based structure of DDT and similar compounds (known as *chlorinated hydrocarbons* or *organochlorides*) is especially dangerous, because it allows these toxic chemicals to insinuate themselves into natural metabolic processes, disrupting them and thereby damaging the very tissues these processes are normally meant to sustain, repair, and defend.

"What sets the new synthetic insecticides apart," wrote Rachel Carson, "is their enormous biological potency. They have immense power not merely to poison but to enter into the most vital processes of the body and change them in sinister and often deadly ways." In humans, Carson argued that this biological potency would manifest itself as toxin-induced cancers and other deadly diseases; in birds, the evidence was already piling up that DDT was threatening some of the most treasured species in the world, including the United States' beloved national bird, the bald eagle.

The 1972 ban against DDT, which resulted from Carson's efforts to raise public awareness about the dangers of the virulent chemical, is counted among the American environmental movement's greatest successes, as it helped to bring the bald eagle and other endangered bird species back from the brink of extinction. There were an estimated 300,000–500,000 bald eagles in the early 18th century; due to DDT contamination and

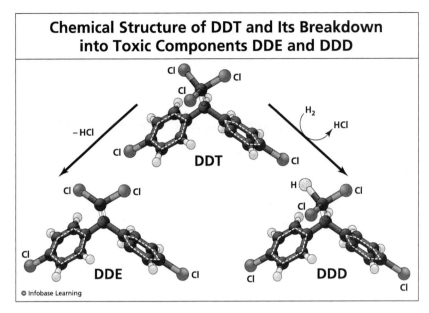

This diagram shows the breakdown of DDT into the toxic components DDE and DDD. All three chemicals are highly persistent in the environment and become magnified as they travel up the food chain.

other threats—including loss of habitat, illegal hunting, power-line electrocutions, and pollution from oil, lead, and mercury—there were only about 412 nesting pairs left in the 48 contiguous states by the 1950s.

Like many other birds of prey, bald eagles sit atop the food chain and suffer the worst effects of DDT and other toxins due to *biomagnification*—a process by which long-lived chemicals become more concentrated as they travel up the food chain. When a predator consumes an animal that has consumed smaller prey or contaminated vegetation, the concentration of contaminants found in the body of the predator becomes greater than the concentration found in prey farther down the food chain. In the case of DDT, the chemical does not kill adult eagles outright, but instead interferes with their ability to metabolize calcium, which in turn affects the bird's fertility or causes it to produce eggs much too brittle to survive.

In 1967, the bald eagle was declared an endangered species in the United States. With regulations finally in place and with DDT banned, the eagle population began to thrive again, and by the early 1980s, the estimated total population had rebounded to 100,000 birds. Their numbers have continued to rise, and by June 2007, the bald eagle was officially removed from the federal list of endangered and threatened species and was assigned a risk level of "Least Concern" on the International Union for Conservation of Nature (IUCN) Red List.

Despite this impressive story of ecological recovery, Carson's warnings about the unknown risks of DDT were hauntingly prescient. Nearly 40 years later, in November 2010, residual DDT has been pegged as a likely new threat to the United States' largest bird, the California condor. Joe Burnett, lead biologist for the central California condor recovery program, found brittle eggshell fragments when he was monitoring an unsuccessful nesting attempt by a condor pair—the first known nesting attempt in the area for at least 100 years.

"The eggshell fragments we found appeared abnormally thin," Mr. Burnett told the *New York Times* in November 2010. "They were so thin that we had to run tests to confirm that it was a condor egg." The shell fragments were reminiscent, he said, of the eggs of birds like brown pelicans and peregrine falcons, which were hit hard by DDT decades earlier but had since recovered.

The suspected route of exposure starts at the bottom of the Palos Verdes Shelf off the coast of California near Los Angeles, where an estimated 1,700 tons (1.5 million kg) of DDT were dumped by the largest manufacturer of the chemical, the Montrose Chemical Corporation, between the late 1950s and early 1970s. (Montrose poured waste containing DDT directly into local sewers, which conveyed that waste to a Los Angeles County waste treatment plant; contaminated wastewater was discharged from the plant directly into the Pacific Ocean.) Now a designated EPA Superfund cleanup site, the Palos Verdes Shelf is adjacent to a breeding ground for sea lions, which eat local fish contaminated with DDT and then migrate up the coast, stopping off at a beach near Big Sur. This sea lion pit stop has become a favorite feeding ground for the area's condors and is the suspected site of their contamination.

Rachel Carson was right, it seems, to warn of the potential long-term effects of releasing untested chemicals into the environment, and the wisdom of her call for oversight—and for the separation of business interests from safety and health

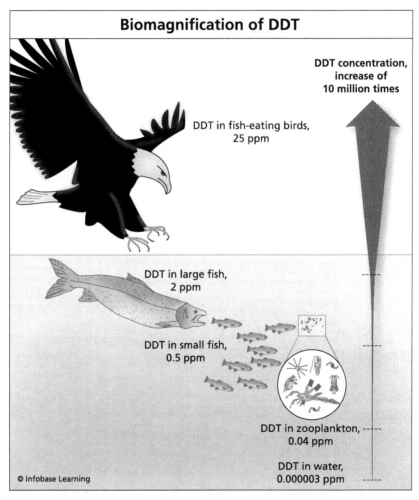

This graphic illustrates the extreme biomagnification of DDT as it climbs the food chain from water, through plankton and fish, into the tissues of fish-eating birds—an increase of approximately 10 million times the original concentration.

Toxic Contamination and Cleanup

A bald eagle guarding its nest *(Tony Campbell, 2010, used under license from Shutterstock, Inc.)*

concerns—extends to all areas of government policy that have an impact on the environment, as can be seen in the recent and tragic example of the massive BP oil spill in the Gulf of Mexico.

THE BP OIL SPILL DISASTER AND ITS AFTERMATH

Around 10:00 P.M. on April 20, 2010, a violent explosion on the *Deepwater Horizon* drilling rig in the Gulf of Mexico killed 11 workers, injured 17 others, and set off the largest accidental marine oil spill in history. At the time of the explosion, the rig was drilling the Macondo exploratory well—named for the Macondo oil and gas prospect in which it was located, approximately 41 miles (66 km) off the Louisiana coast. Nearly three months and several failed containment attempts later, BP Oil, the owner of the well, was finally successful in capping the plume,

Fire response crews battle the blazing remnants of the oil rig *Deepwater Horizon* off the coast of New Orleans, April 2010. *(U.S. Coast Guard photo/Released)*

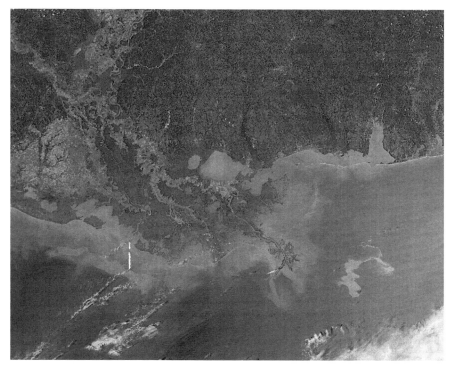

An aerial photo of the BP spill in the Gulf of Mexico, April 25, 2010 *(NASA image courtesy the MODIS Rapid Response Team)*

but not until almost 5 million barrels—185 million gallons (more than 700 million liters)—of crude oil had spewed into the waters of the Gulf. On September 19, a full five months after the explosion, the federal government concluded its own pressure tests of the containment cap, deemed the cement an effective and permanent seal, and declared the well officially dead.

President Obama appointed a special commission to investigate the disaster, and the probe identified multiple faulty decisions and technical failures leading up to the accident—not only on the part of BP but also on the part of Halliburton (the subcontractor responsible for creating and testing the cement mixture that did not properly seal the well) and of Transocean (the owner and operator of the rig). Fred Bartlit, Jr., the lead investigator for the commission, and aides found that Halliburton had conducted three laboratory tests of the cement mixture and that—according to industry standards—the mixture had failed all of them. Moreover, within hours of the explosion, employees of BP and Transocean accepted the results of a pressure test that should have raised serious safety concerns and that pointed to the need for special measures to keep the volatile mix of oil and gas contained—a fateful decision that clearly contributed to the accident.

In the course of its own investigation, BP found that on the morning of the explosion, its team on the rig had decided not to conduct their standard evaluation log for the cement and had broken with their own company's guidelines by moving ahead on the basis of other assessments. "The interesting question is why these experienced men out on that rig talked themselves into believing that this was a good test that indicated well integrity," said Sean Grimsley, a deputy investigator for the commission. "None of them wanted to die or jeopardize their safety. The question is why." This remains an open question in large part, Mr. Bartlit has argued, because Congress did not grant the commission subpoena power, and therefore he and his staff were unable to depose workers under oath in order to resolve contradictory claims made by employees of the three companies.

Ultimately, the commission attributed the accident to not one but a total of nine risky steps BP and its subcontractors took to save time when safer options were available. In its January 2011 report, the panel warned that if significant steps are not taken to rectify systemic breakdowns in communication and oversight and to radically alter the safety culture at oil drilling companies, another terrible accident is likely to occur. "The blowout was not the product of a series of aberrational decisions

---

(opposite page) Many more animals died during the months following the BP oil spill than during the same period in previous years, though researchers are still investigating the causes. This graphic shows where dead or debilitated sea turtles and dolphins were discovered in the months after the spill (as of August 16, 2010). *(Source: New York Times)*

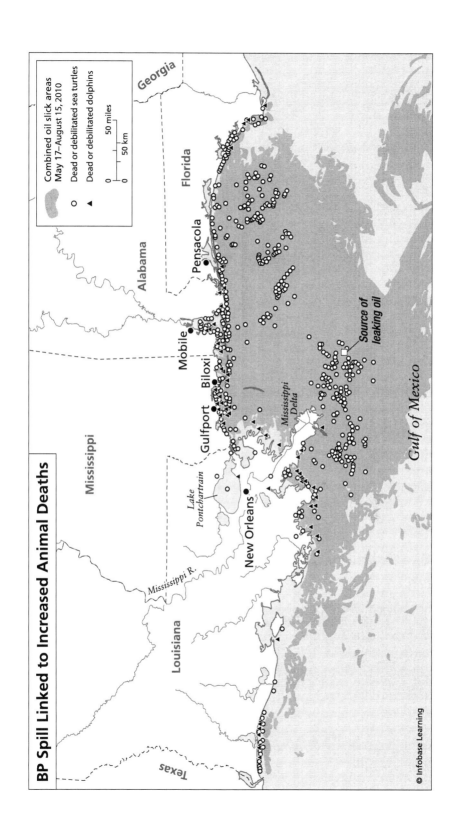

made by rogue industry or government officials that could not have been anticipated or expected to occur again," the panel concluded. "Rather, the root causes are systemic and, absent significant reform in both industry practices and government policies, might well recur."

Perhaps the most tragic finding was that the disaster was totally preventable. Though offshore drilling is an inherently risky business, "the accident of April 20 was avoidable," the panel concluded. "It resulted from clear mistakes made in the first instance by BP, Halliburton and Transocean, and by government officials who, relying too much on industry's assertions of the safety of their operations, failed to create and apply a program of regulatory oversight that would have properly minimized the risk of deepwater drilling."

As of January 2011, the U.S. Department of Justice has filed a civil suit against BP and eight subcontractors for the safety failures leading to the accident. "We will not hesitate to take whatever steps are necessary to hold accountable those who are responsible for this spill," said Attorney General Eric H. Holder, Jr., at a press conference.

Meanwhile, oceanographers and wildlife biologists are tackling the grim work of assessing the short- and long-term environmental damage, both to coastal ecosystems and to deepwater wildlife.

The Extent of the Destruction

The true extent of ecological damage from the massive spill may remain unknown for years—perhaps even decades—but by January 2011 a few initial conclusions were emerging. The first and most encouraging was that less oil had washed onto beaches and marshes than scientists had originally feared. While damage to wildlife, coastal ecosystems, and fishing and tourism industries has been unquestionably severe—and it is critical to stress that assessment of these harms is still ongoing—by these same

(opposite page) The BP spill's underwater effects *(Sources: NOAA, NAS, and Oil Spill Recovery Institute, Louisiana State University)*

The Spill's Underwater Effects

Dispersants on the Gulf's Surface
Chemicals were sprayed onto the slick to speed up the natural dispersion process.

Dispersants were added to the oil causing it to break up into droplets that can be more easily consumed by micro organisms.

Impact on Sea Life

Brown pelicans and other seabirds often dive into the oil because the slick makes the water appear calmer. Coated in oil, they are unable to properly regulate their temperature and can die of hypothermia.

Plankton can be killed by chemically dispersed oil.

All four species of **sea turtles** in the Gulf are already threatened or endangered.

Dolphins have been following response vessels to play, bringing them dangerously close to the oil and chemical dispersants.

Shrimp and other shellfish are especially vulnerable to oil and chemical dispersants because they are stationary.

Fish larvae are most at risk. The Gulf of Mexico is one of only two nurseries in the world for **bluefin tuna.**

Sperm whales spend most of their time diving for prey and may come up directly into the oil when they reach the surface to breathe.

© Infobase Learning

measures of coastal damage, the 1989 *Exxon Valdez* tanker disaster easily remains the most devastating oil spill in history.

The second, much more ominous finding is that at least one—possibly two—enormous underwater plumes of oil from the BP spill did not disperse as quickly as originally hoped and that this underwater oil may threaten deepwater ecosystems for generations to come. In an August 2010 report in the journal *Science,* a team from the Woods Hole Oceanographic Institution (WHOI) led by oceanographer Richard Camilli released the first peer-reviewed assessment of what is believed to be the largest of these underwater plumes. "Our findings," the team wrote, "indicate the presence of a continuous plume of oil, more than 35 kilometers in length, at approximately 1,100 meters depth that persisted for months without substantial biodegradation."

Using a robotic underwater vehicle and a water sampler lowered on a cable, the team tested gulf waters around the BP well from on board the research ship *Endeavor* for a 10-day period in June when the flood from the well was extremely heavy and found that the plume was more than a mile (1.6 km) wide and 600 feet (183 m) deep in places and that it was traveling at a speed of four miles (6.4 km) a day. Based on what his team had recorded in June, Dr. Camilli told the *New York Times* in August that the plume was likely to "persist for quite a while before it finally dissipates or dilutes away."

The WHOI team's findings directly contradicted an early August report by a group of federal agencies and independent scientists led by the National Oceanic and Atmospheric Administration (NOAA). That report had concluded that 74 percent of the oil had been removed or recovered, had evaporated or dissolved, or had dispersed throughout gulf waters as microscopic droplets. The remaining 26 percent, or "residual" oil, the NOAA report concluded, was "either on or just below the surface as light sheen and weathered tar balls, has washed ashore or been collected from the shore, or is buried in sand and sediments." According to that report, the oil that had not yet been removed or recovered—oil in the so-called "residual" and "dispersed" categories, accounting for approximately 50 percent of the spill—was already

(continues on page 64)

TEN EASY WAYS TO CONTRIBUTE TO BETTER WATER QUALITY

In the face of large-scale disasters like the BP spill, it is easy to feel hopeless about the state of the Earth's waters and our ability to protect them. As long as people continue to use petroleum products like oil, gas, and plastics, the risk of massive oil spills will loom large. How can individual actions ever make a difference in the face of corporate greed and widespread consumer apathy? It might be tempting to give up on this particular fight, and direct our time and energy toward other worthy causes.

Not so fast! Individual behaviors can, and do, have a huge impact on protecting local waters—waters that either sustain or harm local communities, animals, and the land on which they live and that feed directly into larger aquatic reserves such as lakes and oceans. Besides working to reduce our use of oil, gas, and plastics, there are several quick and easy steps we can take at home to protect the Earth's waters for generations to come. The following is a list of simple—but extremely effective—ideas:

1. *Store and dispose of household cleaners, chemicals, and hazardous wastes properly, and use nontoxic alternatives whenever possible.* Many common household chemicals are quite toxic and should never be dumped down storm drains, sinks, or toilets. Paints, solvents such as oven cleaners, floor and furniture polish, pool chemicals, insecticides, cleaning products with ammonia and chlorine, antifreeze, and some laundry detergents all fall into this hazardous category. Many communities have special programs and drop sites for disposal of toxic household products like paints, cleaning supplies, insecticides, and batteries. If you are unsure whether yours does, inquire with your local sanitation or environmental health department. Disposing of harmful substances properly once they have been

(continues)

(continued)

used is absolutely critical to keeping our water supply clean; never purchasing or using them is even more effective. Many toxic household chemicals can be replaced by safe and natural alternatives; vinegar and baking soda, for example, are two of the most powerful and versatile household cleaning agents and many paint manufacturers have begun offering low-tox alternatives. For a list of ideas about replacing toxic products in the home, see the EPA's fact sheet on safe alternatives at http://www.epa.gov/region07/citizens/pdf/safercleaning.pdf. For hazardous waste disposal, see the EPA's Web page on household hazardous wastes at http://www.epa.gov/epawaste/conserve/materials/hhw.htm.

2. *Use lawn fertilizers sparingly and natural fertilizers whenever possible.* Contrary to common belief, sometimes fertilizing lawns sparingly—and with the right substances—does more to protect the water supply than not fertilizing at all. If soil is healthy and naturally fertile, no added fertilizer is necessary. When grass is patchy, however, it leaves dirt exposed and more likely to end up in storm drains. Lawn soil can be naturally high in phosphorous (an element that pollutes water by promoting algae growth), and therefore healthy, well-anchored grass actually supports clean, non-polluting runoff. This is why aerating the soil and, if absolutely necessary, using small quantities of fertilizer with low (or no) amounts of added phosphorus, nitrogen, and potassium can support healthy grass while preventing unnecessary mineral pollution. Natural alternatives like compost, mulch, and peat are best. If your family does not already compost, visit the EPA's composting page to learn how: http://www.epa.gov/epawaste/conserve/rrr/composting/index.htm.

3. *Keep storm drains clear of debris.* Never dump waste down storm drains, and keep grass clippings, leaves, brush, and other debris swept away from drain openings. Storm drains connect directly to the nearest streams, rivers, and lakes and can carry leaves, grass clippings, soil, oil, paint, chemicals, and anything else dumped onto the ground straight into local bodies of water.
4. *Wash cars on grass—not driveways—with phosphate-free cleaners.* Washing your car on the lawn rather than on the driveway or street will prevent unnecessary runoff into storm drains, and hand-washing with a bucket and phosphate-free soap—making sure to shut off the hose between rinsings—can save as much as 150 gallons (570 L) of water.
5. *Clean up oil spills and recycle used oil.* Never pour used motor oil into gutters, down storm drains, or onto the ground. According to the Natural Resources Defense Council (NRDC), a single quart of motor oil that seeps into the groundwater can pollute 250,000 gallons (nearly one million L) of drinking water. Your community might already have a program to recycle used motor oil as part of its hazardous waste disposal program; some stores and service stations selling motor oil also have recycling programs.
6. *Landscape the yard in an Earth-friendly way.* Keeping the grass longer—at least three inches (8 cm) high—goes a long way toward keeping rainwater on the lawn and out of storm drains. Using gravel and other porous materials instead of cement; building wood decks instead of concrete patios; and paving walkways with bricks or stone will also prevent unnecessary runoff.
7. *Install a rain garden and/or rain barrel.* When vegetation is planted at elevations lower than paved surfaces, it can catch runoff and allow it to seep slowly into

(continues)

(continued)

the soil (rather than rushing down into storm drains). Rain gardens are depressed areas of land planted with native or deep-rooted vegetation that collects storm water and encourages its absorption into the ground—much like tiny wetland areas—and can be designed quite beautifully. Another great way to catch excess rain is by installing a rain barrel—a holding tank designed to collect or "harvest" rainwater from roofs and terraces for later use.

8. *Dispose of and recycle trash properly.* Be sure never to flush nonbiodegradable objects (like diapers or plastic tampon applicators) down a toilet, as these materials can end up harming sewage treatment systems, polluting waters, and dirtying beaches. Picking up and disposing of pet waste properly helps to keep harmful bacteria out of the water supply.

9. *Do what you can to conserve water.* Many people are unaware that 73 percent of the water used at home gets flushed down the toilet or washed down the shower drain. NRDC suggests using toilet dams or bricks in the toilet tank, which can save four gallons of water per flush, or up to 13,000 gallons (nearly 50,000 L) per year (for a family of four). A dripping faucet wastes as many as 20 gallons (76 L) a day,

(continues)

(opposite page) These figures illustrate the problem of domestic water waste in the United States. The top figure shows that bathroom use represents approximately half of indoor water consumption, while the bottom figure shows that household water use per capita in the United States equals that of Brazil, Germany, the United Kingdom, China, Honduras, and Somalia combined. *(Sources: Mayer et al., 1999, and Pacific Institute)*

Toxic Contamination and Cleanup

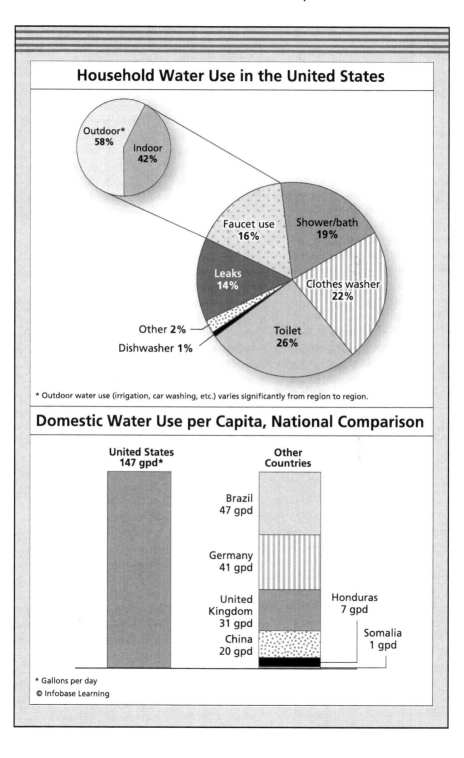

(continued)

and a leaky toilet wastes 200 gallons (760 L). If your family is considering replacing bathroom hardware, low-flow toilets and showerheads are both excellent water conservation options. Other great ways to conserve water around the home include sweeping driveways and sidewalks (instead of spraying them with a hose) and using efficient watering techniques for gardens and lawns. Trickle or "drip" irrigation systems or soaker hoses, for example, are much more efficient than sprinklers or open hoses.

10. *Become an activist!* If your community doesn't have hazardous waste disposal or recycling programs, ask your sanitation, public works, or health department to start one. And there is no need to limit our efforts to local channels; thanks to the Internet, it is becoming much easier to find a collective voice at the national level. Environmental organizations such as NRDC, the World Wildlife Fund (WWF), the Environmental Defense Fund (EDF), the Wilderness Society, and many others are making it quick and easy to communicate concerns about issues like ocean policy, wildlife preservation, toxic contamination, and green energy to government officials through targeted e-mail campaigns. (See the Further Resources section at the end of this book for a more extensive list of national environmental organizations and their Web resources.)

(continued from page 58)

"biodegrading quickly," thanks to oil-metabolizing bacteria that live in abundance in Gulf waters. (For BP's part, its chief executive at the time of the spill, Tony Hayward, went so far in May as to declare that "There aren't any plumes," when explaining why the company was focusing its cleanup efforts on surface slicks.

BP subsequently retracted that claim, accepted responsibility for the disaster, and pledged to create a $20 billion fund to pay for any claims stemming from the accident.)

NOAA's rosy August assessment, comforting as it may have been, was called into serious question by the new study. By measuring the dissolved oxygen in the seawater (which the bacteria would consume as they feed on the oil), the WHOI researchers showed that microbes had barely made headway in breaking down the oil during its first five days in the Gulf. "The idea that 75 percent of the oil is gone and is of no further concern to the environment is just incorrect," Samantha Joye, a professor of marine sciences at the University of Georgia, told the *New York Times*. Dr. Camilli observed that given the very cool waters near the plume (around 40°F, or 4.4°C), the disappointing rate of biodegradation was unsurprising. "In colder environments, microbes operate more slowly," he said. "That's why we have refrigerators."

Also of grave concern in the WHOI findings was the extreme toxicity of the plume, which contained more than 50 micrograms per liter (about 0.05 parts per million) of a group of highly toxic hydrocarbons including benzene, a virulent carcinogen. The team's initial measurements indicated that the oil might be having a harmful effect on bacteria and also on phytoplankton, which serve as a critical food supply for Gulf wildlife.

Ian MacDonald, a Florida State University oceanographer, testified before Congress in August that "I expect the hydrocarbon imprint of the BP discharge will be detectable in the marine environment for the rest of my life. The oil is not gone and is not going away anytime soon."

In November 2010, NOAA released a revised, peer-reviewed estimate largely defending the findings of their original report while acknowledging that some of its assumptions were incorrect. (Chemical dispersants were nearly twice as effective as originally thought, the agency argued, thus compensating for other conclusions that were overly optimistic.)

Jane Lubchenko, administrator of NOAA, cautioned that the new figures were not instructive as to where the residual oil was now or to the spill's ultimate environmental toll. "It does not tell us where the oil is today or its final fate," she said. "This report

Cleanup of the *Exxon Valdez* spill in Prince William Sound, Alaska *(Exxon Valdez Oil Spill Trustee Council/National Oceanic and Atmospheric Administration/Department of Commerce)*

does not address impact at all." The agency also warned that the numbers were rough estimates and subject to revision. As of January 2011, there is still no official estimate of how much oil lingers in the Gulf.

SUMMARY

The infamous 1989 *Exxon Valdez* tanker accident led to new safety regulations for oil tankers (requirements, for example, that tankers be escorted in and out of harbors to prevent them running aground), but these new regulations did not address safety and oversight problems with offshore drilling. The BP disaster drew attention to weaknesses in regulatory oversight of deepwater drilling, and the Obama administration called a halt to nearly all offshore operations until a comprehensive safety review could be completed. In August 2010, the administration promised to overhaul the process by which deepwater drilling permits were granted, requiring more in-depth review of the physical circumstances in each case. In December 2010, the Interior Department said that it would retract an earlier decision to allow expansion of exploratory drilling into the eastern Gulf and along the Atlantic seaboard, saying that no such operations would be permitted for at least seven years. "As a result of the *Deepwater Horizon* oil spill, we learned a number of lessons," Interior Secretary Salazar said when he announced the decision. "Most importantly that we need to proceed with caution and focus on creating a more stringent regulatory regime."

Much in the spirit of Rachel Carson (who, as recounted earlier, called for a separation of the environmental oversight and promotional duties of the USDA), in May 2010 President Obama announced plans to split the Minerals Management Service into two parts—one in charge of safety and environmental oversight and the other responsible for the business end (leasing and revenue collection). Prior to the change, the Minerals Management Service had both reaped the profits from the petroleum industry ($13 billion per year on average) and overseen safety—a system that often allowed the drive for profit to outweigh safety concerns and that permitted petroleum corporations to operate largely unmonitored.

Two otters playing and hugging. Otters are especially vulnerable to the effects of oil spills, as the oil can ruin their pelts' insulating properties and cause them to die of hypothermia. *(Neelsky, 2010, used under license from Shutterstock, Inc.)*

"The job of ensuring energy companies are following the law and protecting the safety of their workers and the environment is a big one," said Interior Secretary Salazar when the plan was announced, "and should be independent from other missions of the agency."

In perhaps the most far-reaching change of all, in July 2010 President Obama signed an executive order creating the first federal policy to clean up and protect the oceans. The order created a National Ocean Council that will oversee any activities having an impact on the sea—activities currently administered by more than 20 federal agencies and governed by more than 140 laws—and it provides for coordinated action plans to more efficiently reduce marine pollution and protect wildlife.

The next chapter focuses on another crisis that the new ocean policy is crafted to address: the overexploitation of endangered species.

4

Species Overexploitation, Protection, and Captive Breeding

A world without tigers, whales, wolves, and giant pandas may seem inconceivable, but it is far from impossible. The rate of species loss due to human activity has reached breathtaking speed in recent centuries, with an average of one species of bird and one species of mammal lost per decade between 1600 and 1700. This already alarming average rose to one bird and one mammal species per year between 1850 and 1950 and then soared to four species of each lost per year in the brief four-year period between 1986 and 1990. As of 2010, one-fifth of all plant and animal species are threatened with extinction, according to the Red List of Threatened Species of the International Union for Conservation of Nature (IUCN).

Species loss can occur when a plant or animal population is harvested or hunted by humans to such a degree that their wild population is threatened or extinguished. Infamous examples include the overhunting of the North American buffalo and, more recently, the gray wolf; the decimation of great whale populations by commercial whaling; the hunting of elephants for ivory; and the poaching of tigers for their bones (for use in traditional Chinese medicine) and for their beautiful pelts. Taking this last tragic example, three out of nine tiger subspecies have gone extinct in the last century alone. Illegal poaching remains a major threat to the long-term survival of the remaining subspecies, in concert with habitat fragmentation and declining populations of their natural prey.

This chapter charts the dire prospects for three of the world's most treasured, most ecologically important, and most hunted large-mammal species—great whales, tigers, and wolves—and the conservation efforts that are working best to protect them.

COMMERCIAL WHALING ENDANGERS MOST OF THE WORLD'S GREAT WHALES

The introduction of factory ships and new whaling technology in the early 20th century made whale hunting far more efficient, and whale catches began to outpace the natural ability of species to replenish themselves. By the middle of the 20th century, many of the great whale populations had been decimated by unrestricted hunting, with a staggering 66,000 whales being killed in a single year (1961).

Several species were nearly extinct by 1986, when the International Whaling Commission (IWC) introduced a *moratorium* on whale hunting in the hope that threatened species would rebound. Animal lovers are sometimes shocked to learn that despite this powerful international agreement, whales are still hunted today. Even with two decades of protection, 10 great whale species are still endangered or threatened, and an estimated 1,000 great whales are killed each year by whaling nations. As if circumstances were not treacherous enough, whales, dolphins, and porpoises also face new and serious threats from entanglement in fishing lines, collisions with ships,

The *Yushin Maru,* a Japanese whaling ship, catches an Antarctic minke whale in the Southern Ocean in December 2001. *(© Jeremy Sutton-Hibbert/Alamy)*

toxic pollution, gas and oil drilling in native feeding grounds, climate change, and overall habitat degradation.

After the IWC moratorium passed, Japan began whaling again under the auspices of scientific research (which is still allowed by the agreement). Japan's deputy whaling commissioner, Joji Morishita, told BBC News in 2008, "The reason for the moratorium [on commercial whaling] was scientific uncertainty about the number of whales. . . . It was a moratorium for the sake of collecting data and that is why we started scientific whaling. We were asked to collect more data."

But the government of Australia, along with many conservation groups, claims that the Japanese "research" is simply a disguise for commercial whaling operations to meet the demand for whale meat, which continues to be a popular part of Japanese cuisine.

Japan is not the only country that continues to hunt whales. Norway registered an objection to the IWC agreement and therefore is not bound by it, and under the agreement several indigenous populations for whom whale hunting is a traditional form of subsistence are also permitted to hunt whales in much smaller numbers.

More than 80 species of whales, dolphins, and porpoises (known as a group as *cetaceans*) inhabit the world's ocean and river ecosystems. Cetaceans, like humans, are warm-blooded mammals that give birth to and nurse live babies, and they are considered second only to humans on many measures of intelligence. "At the very least, you could put it in line with hunting chimps," Hal Whitehead, a biologist and expert on sperm whales, told the *New York Times* in June 2010. "When you compare relative brain size, or levels of self-awareness, sociality, the

---

(opposite page) These graphics show the status of Southern Ocean whales before and after 20th-century whaling. Figure A compares pre-whaling population estimates for several species of whales with the latest available population estimates; figure B shows the number of Southern Ocean whales killed in the 20th century; and figure C shows the number of whales killed by Japan in the Southern Ocean under the scientific research exemption to the international whaling moratorium. *(Source: WWF, 2010)*

Species Overexploitation, Protection, and Captive Breeding

importance of culture, cetaceans come out on most of these measures in the gap between chimps and humans. They fit the philosophical definition of personhood."

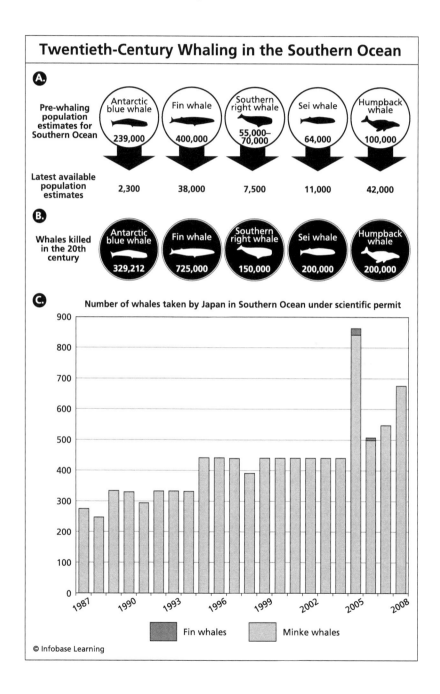

A dolphin drowns when its tail becomes entangled in a fishing net.
(© Ross Davidson/Alamy)

Cetaceans fall into two broad categories: *mysticetes* or baleen whales (also known as great whales) and *odontocetes* or toothed whales, including more than 70 species of whales, dolphins, and porpoises. The accompanying table lists endangered and threatened species from both the baleen and toothed whale groups.

Concerns about the continued vulnerability of whales sparked a controversial proposal to allow the first legal commercial whaling since the 1986 moratorium. The agreement would have allowed some hunting subject to strict yearly limits and close monitoring, and, according to the IWC, the plan would have saved several thousand whales per year. But international talks collapsed in June 2010 when Japan refused to phase out its annual hunt in the Southern Ocean whale sanctuary, where 80 percent of the world's whales go to feed in the summer. Australia is currently suing Japan in the International Court of Justice to put a stop to their whaling operation in the Southern Ocean.

THE STATUS OF THREATENED WHALE SPECIES

Species	Population	Status and Listings*
Baleen whales		
North Atlantic right whale	300–400	endangered (ESA, IUCN)
North Pacific right whale	unknown	endangered (ESA, IUCN)
southern right whale	7,000 +	endangered (ESA)
blue whale	10,000–25,000	endangered (ESA, IUCN)
fin whale	50,000 +	endangered (ESA, IUCN)
sei whale	80,000	endangered (ESA, IUCN)
bowhead whale	7,000–10,000	endangered (ESA)
humpback whale	60,000 +	endangered (ESA)
gray whale	15,000–22,000	endangered (ESA)
Toothed whales, dolphins, and porpoises		
sperm whale	200,000 +	endangered (ESA); vulnerable (IUCN)
killer whale	50,000 +	endangered (ESA)
beluga whale	150,000 +	endangered (ESA)
vaquita	500–600	endangered (ESA, IUCN)
baiji/Chinese river dolphin	functionally extinct	endangered (ESA, IUCN)
South Asian river dolphin (including Ganges and Indus river dolphins)	unknown	endangered (ESA, IUCN)
Hector's dolphin	7,000	endangered (IUCN)
Irrawaddy dolphin	unknown	vulnerable (IUCN)
franciscana	unknown	vulnerable (IUCN)
Atlantic humpbacked dolphin	unknown	vulnerable (IUCN)

* ESA denotes a listing according to the Endangered Species Act; IUCN denotes a listing according to the 2008 IUCN Red List assessment (the latest available).

Sources: IUCN Red List of Threatened Species. Available online. URL: www.iucnredlist.org; NOAA Office of Protected Resources. "Marine Mammal Species Under the Endangered Species Act (ESA)." Available online. URL: www.nmfs.noaa.gov/pr/species/esa/mammals.htm.

PROTECTING THE LAST OF THE BIGGEST BIG CATS

The Chinese Year of the Tiger, 2010, saw the six remaining subspecies of the legendary and beautiful animal on the verge of annihilation. Tiger populations have fallen by 97 percent during the last 100 years, with the remaining 3 percent forced onto ever-shrinking and fragmented scraps of habitat. As few as 3,200 tigers still exist in the wild, and many of these animals now live in small, isolated groups that cannot provide the genetic diversity necessary for the long-term survival of the species.

"Tigers are being persecuted across their range—poisoned, trapped, snared, shot and squeezed out of their homes," said World Wildlife Fund (WWF) tiger advocate Mike Baltzer in a February 2010 press release (as the Year of the Tiger began). Despite their endangered status, tigers are still commonly poached to feed the illegal trade in bones, skin, and other body parts used in traditional Chinese medicinal products such as tiger-bone wine, which has been used for centuries to treat pain and inflammation and to strengthen bones, muscles, and connective tissue.

Strange as it seems, this beautiful animal—an international symbol of raw, untamed power—is now more commonly found in captivity than in the wild. There are at least five times more tigers (an estimated 15,000–20,000) living in captivity than in their natural habitat. China is the largest breeder by far, with an estimated 5,000 to 10,000 animals subsisting on tiger farms where they are bred for slaughter, kept for entertainment, or sold as pets.

"These are speed-breeding factory farms," Conservation International (CI) tiger specialist Judy Mills told BBC News in January 2010. Cubs are taken away from their mothers and suckled by other mammals—pigs or dogs, for example—so that tigresses can bear more cubs at an accelerated rate (three litters per year, or at least three times their natural pace). "The part [of the farm] which people rarely see is basically a winery in which the skeletons of grown tigers are cleaned and put into vats of wine," says Ms. Mills.

China is not alone in breeding tigers in captivity; at least 8,000 tigers live in captivity elsewhere around the world, many

IS CAPTIVE BREEDING AN ANSWER FOR SPECIES DEPLETION?

Xiang Xiang, the first captive giant panda to be released into the wild, was found dead on a forest reserve in China's southwestern Sichuan Province in February 2007, less than a year after his radical relocation. Hopes for a successful life in the wild were high, since Xiang Xiang had received three years of training in survival skills such as den building and foraging for food. Tragically, however, the sweet-faced bear was simply no match for the more aggressive, territorial wild pandas already living on the reserve.

"Xiang Xiang died of serious internal injuries in the left side of his chest and stomach by falling from a high place," said Heng Yi, an official at the China Giant Panda Protection and Research Centre in Wolong where Xiang Xiang was raised. "The scratches and other minor injuries caused by other wild pandas were found on his body. So Xiang Xiang may have fallen from trees when being chased by those pandas."

Despite the high-profile loss, China's captive breeding program has forged ahead, with the aim of increasing the number of giant pandas in the wild (currently hovering at around 1,600). In May 2010, China announced plans to build a new $8.8 million dollar (60 million yuan) captive breeding facility that would take a much more gradual approach to training pandas in survival skills before releasing them into wild.

Sometimes, captive breeding is the only hope for an endangered species. Take the California condor, which was nearing extinction in 1987 when the last 22 wild condors were captured to begin the breeding program. In the early years of the experiment, many captive-bred condors died upon reintroduction to the wild from colliding with power lines and ingesting lead bullets, but since the mid-1990s, the birds have been trained to avoid power lines and their numbers have rebounded.

Other captive-breeding success stories include the black-footed ferret and the Guam rail, a small flightless bird nearly

(continues)

(continued)
eradicated from its native habitat of Guam by the brown tree snake (see chapter 5 for more on this invasive snake). But captive breeding programs cannot succeed without strong measures to monitor and protect animals once they are reintroduced into the wild. One captive-breeding success-story-turned-failure is the reintroduction of the Arabian oryx to Oman. The oryx, a beautiful species of desert antelope, was hunted to extinction in the wild by 1972 but was successfully bred in zoos and then reintroduced to Oman in 1982. Due to illegal poaching and habitat degradation, however, their numbers have plunged yet again from a height of 450 in 1996 to a mere 50 (all males) in 2008.

Other infamous stories of captive-bred animals trying to make it in the wild include the following:

- *Keiko, the orca whale that inspired the movie* Free Willy. He was released into the Icelandic sea in 1998 after years of preparation and training, but instead of bonding with his fellow killer whales, he became a loner who had trouble fishing for himself. He learned to scrape by, scavenging food from Norwegian fishermen and tourists, but he died of pneumonia in 2003.
- *An unwanted pet python released into the Florida Everglades in 2005.* The constrictor attempted to swallow an alligator whole, ultimately exploding and killing both animals.

of them in the United States as pets or in zoos. Weak regulations make accurate counts impossible, but most estimates place the number of captive tigers in Texas alone at more than 4,000—more than wild tiger populations worldwide.

The future for tigers looks bleak, but not futile. Experts say there is time to reverse the trend of destruction and that with the proper protections in place, remaining subspecies could

> ○ *Mara the lioness, aka Elsa, from the popular 1966 film Born Free. She was raised in captivity and released into the wild with her three cubs, and died soon after of a parasitic blood disease. The ultimate fate of the cubs is unknown, though at one point one was found living with two unrelated lions on the Serengeti Plain.*
>
> Unlike mammals—who tend to bond with and become reliant upon their human captors—much simpler animals (fish, for example) are governed by comparably uncomplicated behavior and are better candidates for captive breeding. Some fish species bred in hatcheries have been successfully introduced to depleted wild areas, but biologists have expressed concern that the practice might reduce genetic diversity in wild populations. In October 2007, a study of steelhead trout published in the journal *Science* found that in trout born in hatcheries and reintroduced to Oregon's Hood River, domestication radically reduced their reproductive capabilities (by approximately 40 percent per captive-reared generation).
>
> Given the unknown and potentially dangerous genetic and ecological consequences of reintroducing captive populations to the wild, wildlife biologists caution that captive breeding should be considered a stopgap measure only and emphasize that protection of existing wild species and their habitats should remain the top conservation priority.

thrive throughout their traditional range. Organizations such as WWF, with aid from donors such as Leonardo DiCaprio who donated $1 million in 2010 to save tigers, are working overtime in the fight to protect these magnificent creatures. In a June 2007 paper in the journal *BioScience,* Eric Dinerstein and colleagues cited success stories in the Russian Far East and in Nepal where tiger populations have rebounded from overexploitation, thanks to cooperative efforts between national and local governments and conservation organizations.

A MAJOR SETBACK FOR THE GRAY WOLF, ROCKY MOUNTAIN KEYSTONE SPECIES

In April 2011, congressional leaders agreed to a budget bill rider that would strip the gray wolf of its endangered species status across most of the Northern Rockies. The deal was struck after a federal judge in July 2010 ordered that the critical rocky mountain species be brought back under federal protection in the states of Montana and Idaho. Gray wolves in those states had been removed from the endangered species list under rules crafted by George W. Bush's administration and upheld by President Obama's Interior Department.

First implemented in March 2008, the Bush regulations also removed protections for wolves in Wyoming, but legal action by environmental groups stopped the 2008 fall hunt in all three

FWS biologists take blood samples from a tranquilized gray wolf after fitting it with a radio collar in Yellowstone National Park, Wyoming, in May 2003. *(William Campbell/FWS)*

states. A review by the U.S. Fish and Wildlife Service (FWS) resulted in a new proposal that removed protections for wolves in Montana and Idaho but kept them in place in Wyoming, where the state management plan was deemed inadequate since it allows any wolves—not just the few so-called problem wolves that attack livestock—to be hunted as predators and shot on sight or simply to be shot for sport. (In especially controversial kills, people in Wyoming were observed shooting wolves from airplanes and hunting them down on snowmobiles.)

In a move that shocked and disappointed wolf defenders, the Obama administration chose to go forward with the revised plan in March 2009, and the hunt was on. Defending the administration's decision, Interior Secretary Ken Salazar pointed to the growth of wolf populations since their reintroduction to the Rockies in the mid-1990s. "The recovery of the gray wolf throughout significant portions of its historic range is one of the great success stories of the Endangered Species Act," Salazar said. "Today, we have more than 5,500 wolves, including more than 1,600 in the Rockies." About 250 wolves were killed in the fall 2009 hunt, and environmental groups sued again to restore protections.

In summer 2010, a federal judge in Montana decided in the wolves' favor. Judge Donald W. Molloy ruled that wolves "can be endangered wherever they are within the range" of the distinct population of wolves covered by federal protections, and that for the federal government to protect an endangered species in one state but not others was "like saying an orange is an orange only when it is hanging on a tree." But Judge Molloy's decision was nullified when Representative Mike Simpson (R, ID) and Senator Jon Tester (D, MT) inserted a proposal into the budget bill that would remove gray wolves in Montana and Idaho from the federal list of endangered species. The rider was approved by the leadership of both the House and Senate and consented to by the White House.

Wolf defenders were outraged, arguing that the gray wolf's numbers have not recovered to a sustainable level and that they should remain under federal protection for years to come. Gray wolves are a top predator and a keystone species—a species whose presence maintains the healthy equilibrium between other endemic species. The reintroduction of wolves to Yellowstone has been credited with restoring the health of the entire ecosystem, including several other threatened animal species. (When wolves kill an elk, for example, the wolves consume part of the carcass and then leave the rest to be scavenged by eight other carnivore species—coyotes, bald eagles, golden eagles, grizzly bears, black bears, ravens, magpies, and red foxes—and to a lesser degree by as many as 20 other species.)

Following the fall 2009 wolf kill, Montana's wolf numbers dwindled to just 500 and Idaho's to 835. Before Judge Molloy's ruling stopped the 2010 hunt, both states had increased their kill quotas. "In all the decades of the Endangered Species Act," said Rodger Schlickeisen, president of the environmental group Defenders of Wildlife, "Congress has never legislatively removed protections for any species. It's bad to do it for the wolf, and it could set a very bad precedent, replacing scientific determinations with politics."

SUMMARY

With so many beloved and ecologically essential species hovering on the brink of extinction, it is easy to lose faith in efforts to protect overexploited species. But in this—and in so many other cases presented in this book—knowledge is power. When local, state, and national governments feel adequate pressure from informed citizens and from the international community to take concrete action on behalf of vulnerable species, their dwindling populations can rebound to healthy and sustainable numbers. Effective conservation measures for overexploited species include enforced bans on poaching, severe restriction or elimination of legal hunting, and delineation and protection

---

(opposite page) These maps show the near-elimination of the gray wolf's formerly extensive range and the recovery of extremely limited portions of the wolf's habitat after it was placed on the Endangered Species List in 1974. *(Source: FWS, 2010)*

Decimation and Restoration of the Gray Wolf's Range

Historical Gray Wolf Range

Historical range

Gray Wolf Range at Time of Listing under the ESA (1974)

Range at time of listing

2010, Gray Wolf Range and Southwest Recovery Area

Currently occupied range

Southwest recovery area

© Infobase Learning

of sufficient stretches of habitat to support endangered populations as they recover and grow.

The next chapter looks at a distinctive set of cases in which a particular plant or animal species has not been overexploited at all, but instead is thriving at the expense of other species in an otherwise healthy ecosystem. These are instances in which human intervention is often helpful in curbing so-called invasive species—species not native to a region that are disrupting the delicate balance of the ecosystem and from which humans, the ultimate invasive species, can take critical lessons in how to restrict their own destructive ecological influence.

Invasive Species and Their Impact on Diversity

A large and hungry variety of Asian fish is intentionally introduced to fish farms in the United States in the hope that it will clean up unwanted algae; fast forward a few decades and the White House is pledging nearly $80 million in federal funds to address widespread panic that the fish has found its way into the Great Lakes. If and when that happens, Asian carp could quickly burn through the lakes' food supply, exterminating native species and leaving the vast freshwater ecosystem in ruins.

Invasive species are being introduced into vulnerable ecosystems at a staggering rate—some intentionally, some accidentally—and many with devastating consequences. This chapter begins with two recent, high-profile examples and the damage they have done to local ecosystems: the voracious brown tree snake, introduced accidentally to Guam from the South Pacific and responsible for the obliteration of nearly all the island's native bird species, and Asian carp, introduced intentionally to fish farms in the southern United States before escaping into the Mississippi watershed and now threatening the entire Great Lakes ecosystem—the largest reserve of free-flowing freshwater and freshwater organisms on Earth.

Not all biologists agree that the introduction of invasive species is an ecological change to be feared; some argue that newly introduced species often promote diversity rather than hindering it. The second part of the chapter takes this radically different perspective on invasive species, weighing evidence that in some cases exotic species can be desirable—or at least harmless—additions to ecosystems.

Other cases call for human intervention to restore the health of the affected ecosystem. The chapter's third section tells a satisfying success story of marine biologists, a makeshift contraption called a Super Sucker, and the native coral reefs they are bringing back to life with ingenuity and hard work. Finally, the chapter concludes with a look at how global climate change is turning native species into invasive species at a record pace—with utterly unpredictable implications for life as we know it.

SMALL SNAKE, BIG FISH: INVASIVE ANIMALS WIPE OUT NATIVE SPECIES

Perhaps the most reviled invasive predator of the last half-century, the brown tree snake, has caused the extinction of nearly every native bird species and half the native lizard species on the Pacific island of Guam. When the snake was accidentally introduced to the island after World War II—most likely as a stowaway in military cargo arriving from the South Pacific—it encountered very few natural predators and its population grew unchecked, reaching a current staggering count of up to 13,000 snakes per square mile of jungle. Within just a few decades on the island, the snake had obliterated a total of 10 native bird species, with the remaining two left struggling for survival at a mere 200 members each.

Recent studies suggest that the brown tree snake's impact reaches beyond the devastation of birds and other small vertebrate species to the very health of the forest itself. As seen in chapter 2, birds play a critical role in maintaining forests by dispersing seeds and pollen, thereby promoting diversity

Invasive Species and Their Impact on Diversity

A brown tree snake (*Boiga irregularis*) in Queensland, Australia *(© Ben Nottidge/Alamy)*

of native species throughout forest habitats. The loss of native birds in Guam, biologists say, has severely curtailed reproduction of native tree species and has threatened a variety of plant and animal species that call the forest home.

"The brown tree snake has often been used as a textbook example for the negative impacts of invasive species, but after the loss of birds no one has looked at the snake's indirect effects," said Haldre Rogers, a doctoral student in biology at University of Washington. "It has been 25 years since the birds disappeared. It seems to me the consequences are going to keep reverberating throughout the community if birds are fundamental components of the forest," she said in an August 2008 press release.

Rogers and fellow researchers studied the problem by comparing seed dispersal and seed-coat removal (which aids in germination of trees and is most likely performed by digestive systems of birds) on Guam and on the neighboring island of Saipan. Saipan shares a similar climate and ecology but has, to date, escaped invasion by the tree snake. The researchers placed seed traps at varying distances and found seeds in nearly every trap on Saipan—though more seeds were found in closer traps than in traps farther away—but on Guam they found seeds only in traps directly beneath the canopy of the parent trees. Moreover, most of the farther-dispersed seeds on Saipan had had their seed coats removed (presumably by birds), whereas all of the seeds found on Guam still had intact seed coats.

"It seems logical that if there are no birds then seeds are not able to get away from their parent trees, and that is exactly what our research shows," Rogers said. "The magnitude of difference between seed dispersal on Guam and Saipan is alarming because of its implications for Guam's forests, and for forests worldwide experiencing a decline or complete loss of birds."

In an experimental attempt to curb the snakes' eating rampage, the U.S. Department of Agriculture (USDA) began dropping dead mice laced with Tylenol into Guam's forests in 2010 after research showed that acetaminophen—the active ingredient in Tylenol—is fatal to brown tree snakes in small doses. If deemed successful in controlling local snake populations, the program could be expanded to the entire island within a year.

The U.S. Great Lakes ecosystem has recently been threatened with a similar degree of devastation by a trio of voracious invasive fish—bighead, silver, and grass carp, known collectively as Asian carp after their continent of origin. The carps' assault on the Mississippi watershed is a lesson in how a seemingly good idea can have unforeseen and devastating consequences. The carp were first imported to the United States from China and Siberia in the 1970s to chomp up algae that were choking fish farms in the South, but they did not stay neatly contained on farms; they were soon discovered in the Mississippi River and many of its tributaries. The carps' unanticipated escape is often blamed on flooding, which may have boosted them over containment barriers and into rivers—though their actual route from farms into the wild is unknown.

In recent years, the fish have moved northward to overtake sections of the Mississippi, the Illinois, and other Great Lakes region rivers, hoovering up plankton on which other fish depend and potentially endangering native species (along with the region's $7 billion fishing industry). The large, surprisingly athletic carp have also received negative press for injuring fishermen and recreational boaters. Full-grown silver carp—nicknamed "flying carp" for their ability to leap as high as 10 feet (3 m) into the air—can reach a bruising 100 pounds (45 kg) and have made national news by breaking noses, jaws, even vertebrae of boaters and water-skiers.

Concern over the spread of Asian carp shifted to panic in January 2010 when genetic material from the fish was discovered in a harbor in Lake Michigan and in a river within half a mile of the lake. No live fish were located, but the possibility loomed large that the fish had already breached—or were close to breaching—electric barriers that protect the lake. "It's a big admission of failure," said Henry Henderson, director of the Natural Resources Defense Council's (NRDC) Midwest program. "It indicates the kind of thing we've been fearing since 1993."

Bighead and silver carp jump out of the Illinois River near Havana, Illinois, in May 2007. (© Jason Lindsey/Alamy)

Earlier that day, the U.S. Supreme Court had denied a request for an emergency injunction to force the closure of a Chicago shipping canal that provides direct access to the lake. With the discovery of genetic material from the carp in and around the lake, the governors of Michigan and Wisconsin renewed their call for an emergency shutdown of shipping traffic between Lake Michigan and regional waterways—a proposal that Illinois state officials roundly rejected on the grounds that it would jeopardize Chicago's commercial shipping industry.

In February 2010, U.S. officials pledged $78.5 million to fund efforts to keep the carp out of the Great Lakes, which contain 20 percent of the world's freshwater and no natural predators to keep the carp from annihilating native species. The plan called for new barriers to prevent the fish from entering Lake Michigan via nearby waterways and for the locks that allow ships in and out of the lake to be opened less frequently—a measure that Michigan officials deemed wholly insufficient for preventing an influx of the ravenous fish. "They just need to shut the locks down, at least temporarily," Governor Jennifer Granholm of Michigan told the *New York Times* after the plan was announced.

Five states—Michigan, Minnesota, Ohio, Pennsylvania, and Wisconsin—and a Native American tribe, the Grand Traverse Band of Ottawa and Chippewa Indians, sued in August 2010 in federal court to force closure of Chicago-area shipping locks, but they lost that bid in December. While the federal district judge agreed that ecological harm from the fish could be severe, he found that the states had not provided sufficient evidence of an imminent threat.

COMPETITION FROM EXOTIC SPECIES: FRIEND OR FOE TO DIVERSITY?

Some biologists argue that invasive species, on balance, may have a positive influence on ecosystems: They may promote the growth of diversity rather than deterring it, and in some cases they may actually benefit native flora and fauna rather than threatening them. In an August 2008 paper in the *Proceedings of the National Academy of Sciences* (*PNAS*), ecologist Dov

Sax of Brown University and marine biologist Steven Gaines of University of California, Santa Barbara, analyzed New Zealand's invasion by nonnative plants—most of them transferred intentionally by European settlers—and found that their overall impact was far from negative.

"I hate the 'exotics are evil' bit, because it's so unscientific," Dr. Sax told the *New York Times* in September 2008. When it comes to predicting the threat of a newly introduced organism, it matters most where the species lands on an already-established food chain. Of the extinctions of native vertebrates that have been attributed to invasive species all over the world, Sax and Gaines found that four-fifths of those extinctions had been caused by invasive predators such as foxes, cats, and the brown tree snake discussed in the previous section. "If you can eat something, you can eat it everywhere it lives," said Sax.

When it comes to species that are not predatory, however, Sax and Gaines argue that the effect is much less simplistic. Of the 22,000 nonnative plants brought to New Zealand for agricultural and other reasons, 2,069 have established themselves in the wild, crowding into ecosystems already occupied by 2,065 native species (and officially surpassing endemic plants in terms of total number of species). Rather than dominating the island's ecosystems, they have largely shared them; the island's diversity has effectively doubled with the introduction of those exotic species, with only three native plant species pressured into extinction.

"The overall pattern almost always is that there's some net increase in diversity," said Dr. James Brown, biologist with University of New Mexico and cofounder of the field of *macroecology,* which seeks to understand patterns of diversity and abundance across large geographic scales. "That seems to be because these communities of species don't completely fill all the niches. The exotics can fit in there." He points to the example of Hawaiian waters, where 40 new species of fish coexist with the five original species. The native species have survived despite their tighter living quarters because they are the stronger competitors in highly specialized parts of the niche. These researchers also point out that the introduction of new species can lead to increased diversity through interbreeding with native species and through natural selection in new environments, which can lead to the creation of entirely new species—due either to adaptation of exotic species to their new environments or to increased pressures favoring native individuals with particular traits.

Critics of this kinder, gentler depiction of invasive species, however, insist that it is much too early to judge the ecological impact of new species at the current speed with which they are being introduced. "What's happening now is a major form of global change," McGill University biologist Anthony Ricciardi said in response to Sax and Gaines's research. "Invasions and extinctions have always been around, but under human influ-

---

CUTTING DOWN TREES TO RESTORE A FOREST

Counterintuitive as it may seem, sometimes the best way to restore a forest is to cut down the trees. In a preserve in the middle of New York City, for example, a community restoration group has cut down most of the existing trees and replaced them with native species like oak, gum, and dogwood. "You have to go back to ground zero before you can build it up again." Bruce Stuart, a retired firefighter and longtime resident of the neighborhood, told the *New York Times* in May 2008 that he has seen "an explosion of invasive species" on the preserve during his lifetime. This is especially true of the Norway maples, which have tended to win out over native trees in the life-and-death competition for sunlight and water.

This rather radical approach—eliminating invasive species to allow native species to reestablish themselves—has not been limited to forests. Across the United States, for example, local officials are teaming up with ranchers to use sheep grazing on grasslands as a natural, chemical-

ence species are being transported faster than ever before and to remote areas they could never reach. You couldn't get 35 European mammals in New Zealand by natural mechanisms. They couldn't jump from one end of the world to another by themselves."

There are also clear exceptions to the predator-competitor distinction, as especially aggressive—but nonpredatory—species can also wreak total devastation on ecosystems. Cogongrass, for example, is a weed that was used as packing material on Japanese ships that arrived in Alabama in the early 20th century. Also known as the perfect weed, it has spread swiftly throughout the southeastern United States, killing crops and squeezing native wild plants out of their niches. It is hardy and

---

free way to control invasive weeds and to restore native plant species.

Conservation biologists caution that restoration groups must look at the whole ecosystem before attempting to eliminate a single invasive species, as intervention of any kind can have unanticipated consequences. On Macquarie Island south of Australia, for example, scientists killed off a nonnative species of cat that had been preying on native birds since the early 1800s, only to find that the cats had also been controlling a nonnative species of rabbit. With the cats gone, the rabbits bred unchecked and ravaged the island's lush vegetation, leaving entire hillsides bare and vulnerable to takeover by invasive plant species.

"There have been hundreds of invasive species eradication efforts," said ecologist Erika Zavaleta of University of California, Santa Cruz, "and the vast majority have resulted in clear conservation gains. But Macquarie Island is a new, clear example of unexpected side effects that can happen.... Scientists need to ask themselves key questions, like how all the species on the island interact with each other."

Eradicating cogongrass in Alabama *(Nicole Bengiveno/New York Times/ Redux Pictures Archive)*

tough to kill, and its texture is so rough and unappetizing that animals refuse to eat it. Alabama forester Stephen Pecot told the *New York Times* in September 2009, "People think this is just a grass. They don't understand that cogongrass can replace an entire ecosystem."

Asian carp, discussed in the last section, also fall into this harmful competitor category; they are not endangering other fish by eating them, but by radically depleting their food supply. "If you pour on more species," said Ricciardi, "you don't just increase the probability that one is going to arrive that's going to have a high impact. You also get the possibility of some species that triggers a change in the rules of existence."

Dr. Ricciardi also points to the unknown consequences of combining imported species with climate change and other extreme human impacts. "Invasions will interact with climate change and habitat loss. We're going to see some unanticipated synergies."

Climate change—and the resulting shifts in ecosystems—will be addressed in the final section of this chapter.

VACUUMING THE REEF

Throughout marine environments, fast-growing alien algal species are posing a deadly threat to precious coral ecosystems by growing on top of the reefs, killing them, and disrupting the habitats on which so many other marine species depend. In the Hawaiian Islands, for example, alien algal species are ravaging coral ecosystems, smothering the coral itself and filling in nooks and crannies in the normally complex reef structure that form homes and hunting grounds for so many other species.

"Alien algae is the worst invasive species problem we're dealing with in the state of Hawaii," says aquatic biologist Tony Montgomery. "It likes to overgrow coral and impact the coral communities." To combat the algal crisis, researchers with the Nature Conservancy, the state of Hawaii, and the University of Hawaii have developed a device they call the Super Sucker, a large barge equipped with a suction tube that can remove on the order of 3,000 pounds (1,360 kg) of alien algae per day. Divers hand-pluck algae from the surface of the coral—gently and painstakingly, so as not to damage the delicate reefs—and stuff the offending species into the vacuum tube, which then carries the algae to the surface and deposits it on a sorting table on the barge. Crew members dig through the algae by hand, separating any native plants or animals that may have accidentally been removed and returning them to the reef. The algae is then bagged and distributed to local farmers for use as compost.

The Super Sucker was first used on a section of reef within a protected marine reserve, where a particularly aggressive algal species had taken over and was threatening to spread to other reefs. "We used the Super Sucker on this reef over the course of several months to remove over twenty thousand pounds of alien algae from the reef," said Eric Conklin, marine science adviser for the Nature Conservancy. "After we removed it, quite frankly I expected it to grow back, and I thought we'd be documenting how long it takes for algae to grow back following a removal." But instead of growing back, the algae that was left slowly disappeared, and the reef has stayed algae-free without further intervention for three years. Project members speculate that the operation's long-lasting success is thanks to an abundance of herbivorous fish living a protected life on the reserve. The divers had evidently removed enough algae to reach a threshold level below which the fish could munch up the remaining algae on their own.

More funding is needed, crew members told a video journalist with the *New York Times,* to expand the successful operation to other reefs. "What the Super Sucker does," Conklin observed, "is it buys us time to develop these longer-term management strategies for our oceans, which aren't just about stopping alien algae. They're about restoring health to Hawaii's reefs."

CLIMATE CHANGE ENCOURAGES THE SPREAD OF INVASIVE SPECIES

The world is warming rapidly, with projected average temperatures reaching 2.5 to 10 degrees Fahrenheit above present levels by the year 2100 if atmospheric concentrations of carbon dioxide and other greenhouse gases from smokestacks and automobiles continue to mount. By sobering contrast, it took thousands of years—since the coldest extremes of the last major freeze 17,000 to 21,000 years ago—for the globe to warm five to 10 degrees Farenheit to the levels recorded in the mid-20th century.

In Yellowstone National Park, the effects of climate change are evident in the spread of an invasive plant called the Canada thistle, which has thrived in the warmer, drier weather of the last decade and has forced other plants from their native niches, thus shifting the balance between animal species that depend on the plants. Dr. Robert Crabtree, chief scientist with the Yellowstone Ecological Research Center, told the *New York Times* in March 2008 that as climate change tips the delicate equilibrium in ecosystems like Yellowstone's, "the winners are going to be the adaptive foragers, like grizzlies that eat everything from ants to moose, and the losers are going to be specialized species that

---

(opposite page) This map shows the known global distribution of invasive marine species as of 2008. *(Sources: B. S. Halpern et al., 2008, and* New York Times*)*

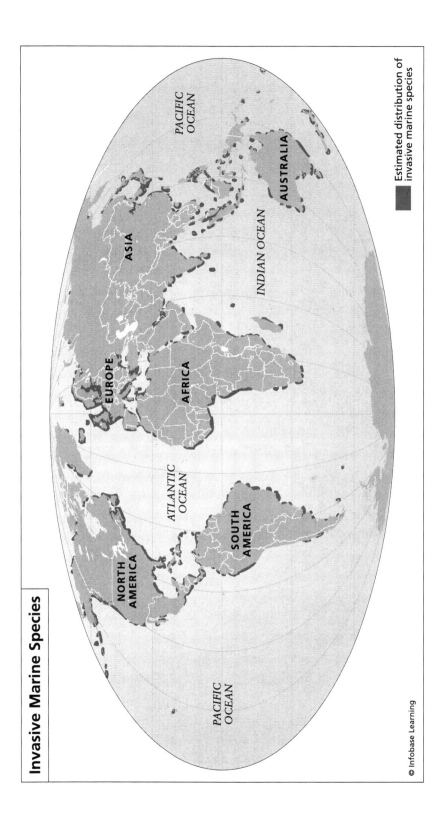

can't adapt," such as the long-tailed weasel and the vole, both on the decline in the park.

There have been other abrupt changes in climate since the last ice age, but as Penn State climate expert Richard Alley told the *Times* in January 2003, "The preindustrial migrations were made without having to worry about cornfields, parking lots and interstates." Dr. Alley expressed particular concern that interdependent plant and animal species are likely to respond to climate pressures by migrating at different rates. To survive,

ANDi, the first genetically modified rhesus monkey, at the Oregon Regional Primate Research Center in 2000. The ultimate influence of genetically modified plants, animals, and microorganisms on global genetic diversity is a wholly unknown variable in the quest to preserve Earth's remaining stores of biodiversity. (© *Mike Stewart/Sygma/CORBIS*)

species will need to change what they eat, "or rely on fewer things to eat, or travel farther to eat, all of which have costs."

In January 2003, a study published in the journal *Nature* found that on average species' ranges are moving toward the poles at some four miles per decade and that spring reproductive events, such as flowering and egg laying, are already shifting 2.3 days earlier per decade. Moreover, the researchers calculated a 95 percent chance that many of these ecological changes are due to climate change rather than to some other factor. "You're seeing the impact of climate on natural systems now," said Wesleyan University economist Gary Yohe, one of the study's authors. "It's really important to take that seriously."

In the same issue of *Nature,* a second group of researchers also reported "a consistent temperature-related shift, or 'fingerprint'" of climate change on a wide range of species based on a meta-analysis of 143 ecological studies. "If we're already seeing such dramatic changes" among species, said the study's lead author, Stanford ecologist Terry Root, "it's really pretty frightening to think what we might see in the next 100 years."

As global temperatures rise, some hardy invasive plant species might turn out to be humans' best horticultural allies as we scramble to adapt to changing climate patterns. Weeds, by definition, are the pests of the plant world, but their relative hardiness and genetic diversity will tend to work to their advantage in rapidly changing temperature and moisture patterns, while many food crops and highly specialized native plant species will struggle and die out—as the grizzlies of Yellowstone have already been forced to discover.

SUMMARY

Modern humans move about the planet with a speed and ease unlike any other species in Earth's history. One unintended consequence of this accelerated mobility has been the destruction of native species—sometimes entire ecosystems—by aggressive exotic species. While serious measures to prevent the accidental spread of disruptive predators and competitors can go a long way toward protecting the health of ecosystems—the successful effort to prevent the brown tree snake from spreading to other Pacific islands is a prime example—as long as humans are moving plants and animals about the planet for agricultural and economic reasons, the lives of native species will be jeopardized in largely unpredictable ways.

The next and final chapter will highlight a single, powerful solution that may address many of the toughest challenges to biodiversity considered in these pages: habitat loss and fragmentation, toxic pollution, overexploitation, and the accidental or intentional introduction of harmful nonnative species. Mega-reserves—or large stretches of land and water sheltered from human influence—may be the best hope for saving endangered species from extinction and for maintaining the complexity of natural ecological processes at a large enough scale to preserve the overall health of ecosystems. Cordoning off mega-reserves will mean inconveniences for humans who are accustomed to roaming freely and using the planet's resources at whim, but it may be the only effective way to protect the natural wealth of the planet for generations to come.

Ecosystem Disruption and the Call for Mega-reserves

When keystone species in an ecosystem are harmed or exterminated by habitat loss, toxic pollution, or other destructive forces, a range of species that depend on them suffers and the fragile web of life can collapse altogether. Coral reefs, for example, are placed at increasing peril as shark populations dwindle, and the acacia tree—a vital player in the vulnerable ecosystem of the African savanna—has shocked wildlife biologists by growing sick when elephants and other large herbivores that eat it are removed. This chapter considers these two examples of ecosystem disruption, both of which reveal that seemingly simple relationships among organisms can in fact be so complex that the consequences of altering those relationships are unknowable until the ecosystem has already been disrupted—sometimes irrevocably.

Aldo Leopold, whose radical ideas about expanding ethical consideration to the entire natural world are summarized in chapter 1, famously noted that "The first rule of intelligent tinkering is to save all the parts." In the continuing effort to defend the planet's remaining stores of biodiversity, not only does it seem that Leopold's assessment was on the mark, but it also appears that the only way to "save all the parts" is to save entire ecosystems by setting aside reserves of land and water large enough to shelter all the systems' parts in a relatively undisturbed state.

WHY CORAL REEFS NEED SHARKS

In the fight to preserve one of the globe's most precious and threatened reserves of biodiversity—its coral reefs—a little fear apparently goes a long way. By comparing the few remaining pristine reefs to reefs that have been disrupted by overfishing and pollution, scientists have discovered that sharks, among the Earth's most intriguing and terrifying creatures, are absolutely

A blacktip reef shark (Carcharhinus melanopterus) swimming over a tropical coral reef *(Ian Scott, 2010, used under license from Shutterstock, Inc.)*

critical to maintaining the health of coral ecosystems. Where sharks and other top predators are patrolling the reefs and keeping smaller predators in check, the corals are clean, healthy, and populated by a greater diversity of creatures—though most of these animals are invisible, because they are hiding.

Dr. Enric Sala, ecologist and coauthor of a comparative study of reefs in a Pacific island chain south of Hawaii, told the *New York Times* in February 2008, "When people see photos they say, 'Well, the water is empty.' For me, it's prettier because the corals are healthy and clean and you don't see seaweed in the reefs and you see these big snappers and sharks." The degraded reefs host more fish overall, but they are much smaller than the fish populating healthy reefs. "The percentage of the bottom cover by large corals declines, the seaweed takes over, then the microbes become much more abundant," Dr. Sala said.

This is so because sharks and other top predators eat smaller predators who—when left to dominate—gobble up the even smaller fish who graze on the algae and keep the coral clean. The ecology in the uninhabited atolls "is similar to what we see in Yellowstone—the landscape of fear," Dr. Sala said. "In Yellowstone there are all these wolves, and the deer are much more attentive." In less healthy reefs, the researchers reported in February 2008 in the journal *PLoS One,* large top-predator species are "virtually absent," a pattern of disruption attributed to overfishing and pollution that Dr. Sala and colleagues wrote "tends to disproportionately reduce densities of longer-lived, larger-bodied individuals." They found that while sharks accounted for 74 percent and 57 percent of the top predator biomass at the two most pristine reefs, they were hardly present at all at the two reefs most disturbed by humans.

These findings have important implications for protecting reefs from immediate dangers like overfishing and pollution and also from longer-term threats like global climate change. Not only were the coral in the uninhabited atolls healthier and the biota more diverse, but the reefs also exhibited greater resilience to major temporary stresses like coral disease and bleaching than did reefs where food chains had been disrupted. This discovery led the researchers to conclude that "local protection from overfishing and pollution may enhance ecosystem resilience to warm episodes and coral bleaching that result from global warming."

Dr. Sala likened the reefs to ecological machines whose parts include a variety of plants, large and small fish, corals, and microbes. "You can hit the system with a disturbance, but the system comes back," he said. If, however, you remove a critical component of the system—top predators, for example—then "the machine is going to malfunction."

When ecological machines like coral reefs are sheltered from these major breakdowns, the benefits can extend well beyond the boundaries of the protected ecosystem. Glover's Reef, a six-island atoll off the coast of Belize, is a heartening example of how marine reserves where fishing is forbidden—so-called no-take reserves—can protect members of coral ecosystems from the tiniest microbe all the way up to top predators like sharks and can also repopulate fisheries outside the no-take zone, thus benefiting local fishermen who depend on the oceans for their livelihood.

"I think Glover's Reef is a model of hope," says Ellen Pikitch, a marine biologist at Stony Brook University who heads up the largest shark population study in the Caribbean. While shark numbers all over the world have fallen sharply in recent years, shark populations at Glover's Reef have remained healthy. The reserve at Glover's Reef, Dr. Pikitch says, shows that "marine reserves, even small marine reserves, can work" and can even do what she dubs "double duty" by restocking local fisheries on which so many people depend.

The no-take model has also proven effective in places like Australia's Great Barrier Reef Marine Park, where a patchwork of hands-off reserves was delineated in 2004. Underwater imaging revealed that two years after the no-take reserves were created, populations of coral trout—a species prized by fishermen—had increased by at least 60 percent. Since larvae can be transported great distances by currents, a study published in the June 24, 2008, issue of *Current Biology* concluded that the no-take reserves will help repopulate coral trout communities in the other two-thirds of the reef where fishing is still permitted.

Based on successes like these in Belize, Australia, and other major fishing cultures such as Micronesia, marine biologists all over the world are calling for expansion of protected ocean

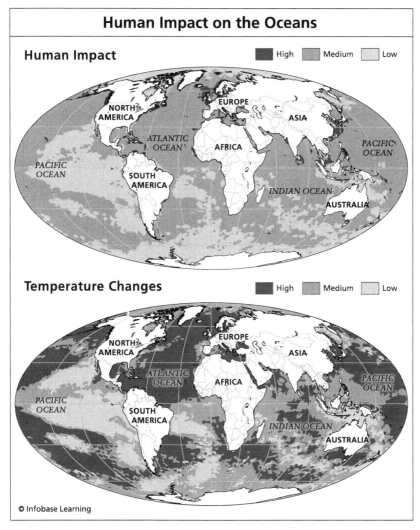

These maps illustrate the known extent of the human impact on oceans, including the combined effects of shipping, fishing, pollution, invasive species, temperature changes, ultraviolet light changes, and acidification due to CO^2 absorption (top figure); and increases in warm temperature anomalies at the ocean's surface over the past two decades (bottom figure). *(Sources: B. S. Halpern et al., 2008, and* New York Times*)*

areas. Dr. Sala, for one, argues that to be effective, these protected marine reserves will need to be "large enough to include healthy populations of top predators." In June 2010, more than 250 scientists from 39 countries asked world leaders to designate many such large marine no-take reserves as part of the Pew Environment Group's Global Ocean Legacy project.

"We have to work on very large trans-boundary marine protected areas with intergovernmental agreements," said one of the statement's cosigners, noted coral reef scientist Bernard Salvat of the University of Paris. "We now need to speak out to educate governments and the public about the crisis facing our oceans and the long-term benefits of establishing large, no-take marine reserves."

ELEPHANTS, ANTS, AND ACACIA TREES: THE IMPORTANCE OF BEING EATEN

Acacia trees are vital components of African savanna ecology, feeding a range of herbivores, serving as nests for insects, and storing carbon and moisture. One species of acacia, the whistling thorn tree, has a *mutualistic* relationship with friendly ants that nest in its thorns and protect it from overharvesting by elephants (by crawling up the large mammals' trunks and biting them). But zoologist Todd Palmer of the University of Florida was intrigued when he noticed that acacias on his research site in Kenya began to sicken and die after they had been fenced off from wild herbivores such as elephants and giraffes. The fenced trees, in fact, were twice as likely to die off as the unfenced ones, and their growth slowed by 65 percent.

"That struck me as paradoxical," Palmer told the *New York Times* in January 2008. "If you remove large herbivores, you should see more vigorous trees." Palmer observed that the sick trees, which had been surrounded by an 8,000-volt electric fence for a decade, had fewer thorn nests. Probing further, Palmer and his colleagues discovered that when threats from large herbivores were removed, acacia trees recognized almost instantly that they did not need their ant partners to protect them from being munched, and they cut back abruptly on their thorn and nectar production. The resulting decline in helpful ants allowed for a takeover by a much less friendly ant that did not need the

Ecosystem Disruption and the Call for Mega-reserves

thorns or nectar and that, in turn, encouraged infestation by an especially destructive insect. "The cavity-nesting antagonistic ants actually promote the activities of the stem-boring beetle," Palmer's coauthor, Stanford biologist Robert Pringle, told *Scientific American* when the study was released. The cavities created by the stem-boring beetle serve as homes for the new ant communities; they also weaken and sicken the trees.

"If you get rid of the large mammals, it shifts the balance of power, because the trees default on their end of the bargain," Palmer said. "When the trees opt out, their hard-working employees starve and grow weak, which causes them to lose out. So, ironically, getting rid of the mammals causes individual trees to grow more slowly and die younger."

It is a balance of power that took millennia to evolve and just a few years to unravel. Dr. Palmer and his colleagues were shocked by how quickly the system fell apart as soon as one

An elephant feeding on an acacia tree in Kenya (© *Eric and David Hosking/CORBIS*)

critical component was removed. It would presumably take the trees many more thousands of years to adapt to life without large herbivores—a grace period that the destructive insects would be unlikely to grant.

DEMOLITION OF ELEPHANT CULTURE AND ITS BRUTAL CONSEQUENCES

Elephants are among the Earth's most social creatures, and to biologists who closely observe their culture they sometimes seem more "human" than humans. For the entire history of their kind up until recently, elephant mothers have raised their young in tight-knit extended social networks that include grandmothers, aunts, cousins, and friends. Young elephants spend the first eight years of their lives within 15 feet of their mothers before being socialized into male and female networks, and elder males perform the critical role of keeping adolescent power-plays in check until the younger males have matured to adulthood.

When an elephant dies, family members enter an elaborate period of mourning, holding vigil over the body for days and then returning regularly for years afterward to sit with the bones, caressing them with their trunks. Elephants are famous for their long memory, and unsurprisingly, recent brain scans performed on elephants have found that they boast an enormous *hippocampus,* the area of the brain known in humans as the major seat of episodic memory (memory of people, places, and events) and a critical hub in the neural circuitry of emotion. Nowhere is this cognitive and emotional sophistication more evident than in the rituals elephants hold around the loss of a loved one.

What happens to these stable, nurturing communities when key members of the herd are killed or relocated? Researchers have discovered tragic consequences as widespread poaching, habitat degradation, and relocation have become the norm in recent years. Many herds now live without male and female elders and without extended child-rearing groups, and the con-

It was not news to biologists that elephants support plant and animal diversity by germinating plants and by opening areas of forest canopy, thus promoting the growth of smaller plants that benefit a variety of animals. What was news was

> sequences are similarly brutal to those seen in human communities when families are torn apart by violent death, illness, crime, and poverty. Elephant attacks on humans have become common in recent years, so much so that a new statistical category—human-elephant conflict (HEC)—was created in the mid-1990s to monitor the growing problem. In Africa, conflicts with elephants have become a nearly daily occurrence, and in parts of Africa and India hundreds of people and elephants have been killed in such conflicts since the year 2000.
>
> In one national park in South Africa, as many as 90 percent of male elephant deaths are now caused by other male elephants (in contrast to only 6 percent in more stable elephant communities). And on a number of reserves in South Africa, the surge of violence has not been limited to humans and other elephants. In a singularly abnormal and cruel development, certain adolescent male elephants have been raping and killing rhinoceroses. It turns out that in all cases of such violence, the young elephants who harmed rhinos had suffered extraordinarily violent childhoods themselves. They had all witnessed members of their families being shot, and in some especially horrific cases, they had even been tethered to the bodies of dead or dying relatives while awaiting relocation.
>
> "The loss of elephant elders," psychologist Gay Bradshaw told the *New York Times* in October 2006, "and the traumatic experience of witnessing the massacres of their family, impairs normal brain and behavior development in young elephants." This finding is unsurprising in light of evidence that early violence and separation from caregivers have been shown to impair brain and behavioral development in another especially social youngster—the human child.

their enormous contribution to the protection of a critical species that they were thought only to harm. "Large herbivores are tremendously important players in these systems," Pringle says. "Not just because of the direct effects they have upon plants, but also because of the myriad effects they exert on smaller, less conspicuous components of biodiversity." And as Palmer notes, "It's becoming increasingly clear that anthropogenic change can have rapid and unanticipated consequences for cooperative species interactions, and we caught this happening in real time."

Research on large herbivores' contributions to ecosystem health is more critical than ever, as African national parks are experiencing ever-dwindling numbers of large mammals. According to a July 2010 survey published in the journal *Biological Conservation,* the past 35 years have seen a decline of nearly 60 percent in large mammals such as lions, elephants, giraffes, buffalo, and zebras due to habitat degradation, insufficient resources and personnel, and a growing bush meat trade.

MEGA-RESERVES: THE BEST FIX FOR ECOSYSTEM DISRUPTION

The urgent need to protect marine ecosystems from overfishing is a clear and compelling reason to create ocean reserves large enough to maintain connected, high-diversity plant and animal communities that include top predators. Successful mega-reserves like those in Belize's Glover's Reef and Australia's Great Barrier Reef bolster the case for expanding protected areas in sensitive marine environments.

The recent movement toward larger-scale marine reserves—both to protect pristine ocean ecosystems and to restore degraded ones—begs the question of whether the current patchwork of small land-based habitat preserves is sufficient to protect rapidly vanishing terrestrial species or whether larger stretches of sheltered areas are essential to preserving land-based biodiversity. Since the early conservation days of Gilbert Pinchot (see chapter 1 for a discussion of his preservation philosophy), the standard wildlife management approach has been to identify small, key patches of habitat and bring them under government control as national parks. This system was politically practical since it required relatively little sacrifice from human neighbors in terms of living space and access to natural resources, but as biologists continue to evaluate the pressures that still threaten protected

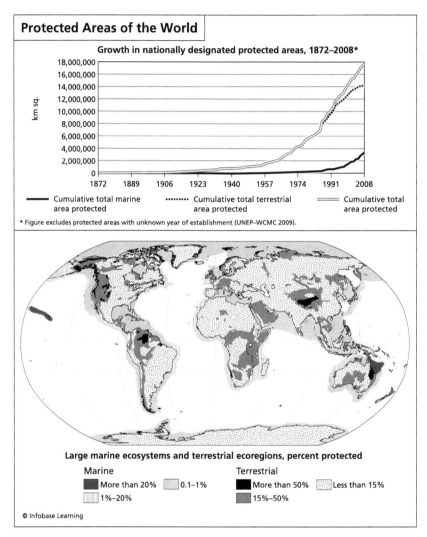

The top figure shows the sharp increase in nationally designated protected areas since 1872, and the bottom figure illustrates the percentage of terrestrial and marine areas that are now protected by national governments. *(Source: UNEP World Database on Protected Areas)*

species—pressures such as habitat degradation and fragmentation, toxic pollution, forced relocation, illegal poaching, global climate change, and the intentional and accidental introduction of invasive species—it is becoming increasingly clear that a radical reenvisioning of our conservation approach is the last, best hope for defending the planet's remaining stores of biodiversity.

Above all, a successful long-term conservation strategy must be built on the knowledge that individual species do not live and die in isolation, but rather that they are connected within a complex web of existence to all other members of their ecosystem. These relationships are often extraordinarily complex or unknown, and therefore the consequences of altering them are largely unpredictable. Given such tremendous uncertainty, the only reliable means of protecting individual species over the long haul is to protect the health of the entire ecosystem in which they are embedded. Coral reefs need sharks, acacias need elephants—and humans were perfectly unaware of these complex dependencies until after they were disrupted.

In their 1999 book, *Continental Conservation,* biologists Michael Soulè and John Terborgh called for the creation of land-based mega-reserves, arguing that without them the "absence of top predators appears to lead inexorably to ecosystem simplification accompanied by a rush of extinctions." Few such terrestrial mega-reserves have been created to date, but one vast reserve in South Africa serves as a shining example for future efforts. In the Eastern Cape, the Baviaanskloof—or Valley of the Baboons—Mega-reserve covers approximately 200 square kilometers of breathtaking mountainous landscape and contains seven of eight of South Africa's *biomes* (distinct ecological communities coexisting in a particular climate). The gorgeous preserve provides a habitat for more than 50 mammal species, more than 300 bird species, and more than 1,000 plant species, while still allowing visits from tourists under highly controlled conditions.

The precious and endangered Amazon region is a natural choice for larger land reserves, and in separate papers, Carlos Peres of the University of East Anglia's Center for Ecology, Evolution, and Conservation and William Laurance of the Smithsonian Tropical Research Institute offer several arguments for the creation of mega-reserves in Amazonia. Small reserves are wholly insufficient first and foremost, they argue, because biologists are still profoundly ignorant of the actual number and types of plant and animal species existing in hard-to-penetrate areas of the Amazon (surveys routinely discover vast numbers of new species), and therefore smaller conservation priorities are being chosen essentially in the dark, perhaps neglecting the region's richest troves of biodiversity.

"Indicators of our present degree of biological ignorance," Peres observed in a June 2005 issue of *Conservation Biology*, "suggest that we have barely dented the surface of a systematic inventory of the Amazonian biota." Without an accurate assessment of the distribution of life in the Amazon, accurate prioritization of smaller conservation hot spots is simply impossible. Laurance echoed this assessment in the December 2005 issue of *TRENDS in Ecology and Evolution*, writing that "In the face of such daunting uncertainty, an expansive network of functionally interconnected reserves is an effective way to capture much of the biodiversity of the region."

Other compelling reasons for Amazonian mega-reserves include the large land requirements for top predators and species migration; the relative resilience of larger stretches of undisturbed land to forest fires, deforestation-induced changes in rainfall patterns, and short- and long-term effects of global climate change; and the lower economic cost of operating fewer, larger reserves (rather than many smaller ones). "For the rapidly disappearing Amazon," Laurance wrote, "the best conservation strategy is to move fast—and think big." This sensibly ambitious strategy looks to hold true for all the planet's most threatened bastions of biodiversity.

SUMMARY

The past six chapters have surveyed some incredible highs and lows in the battle to protect the planet's remaining species from seemingly unmanageable human-made threats. The encouraging news is that a growing number of dedicated scientists, policy makers, and informed citizens are answering the challenge with some truly innovative and effective solutions. However, the critical question remains: What will it take to consistently convert today's gloomy ecological circumstances into tomorrow's lasting successes? Looking to history for guidance, traditional approaches to species conservation have been shaped largely by short-term economic interests and by the outmoded, but deeply rooted, perception that the natural world is to be valued only as a collection of material possessions for future human exploitation and aesthetic appreciation. These traditional approaches have utterly failed to protect much of the natural world from misuse and destruction, and it has become clear that a much more expansive and optimistic approach is called for—one that moves fast and thinks big.

Aldo Leopold pointed the way a half-century ago with his concept of a "land ethic," which, he wrote, "changes the role of Homo sapiens from conqueror of the land-community to plain member and citizen of it. It implies respect for its fellow-members, and also respect for the community as such." Just as Leopold predicted, it is at the level of community—not at the level of individual actors or isolated groups—that the planet's priceless treasures can be effectively defended.

In practice, such a potent environmental ethic means protecting expanses of land and ocean large enough to preserve entire ecosystems from disruption and irrevocable harm. This ambitious—but essential—call to action will only succeed if new generations of researchers, activists, and policy makers see themselves as vital organs within a vast, complex living system, organs that depend upon, and are critically important to, the well-being of the system as a whole.

CHRONOLOGY

1789 Jeremy Bentham is the first utilitarian philosopher to argue against the infliction of suffering on animals: "The question is not, Can they *reason?*" Bentham writes, "nor, Can they *talk?* but, Can they *suffer?*"

1914 Gifford Pinchot, first chief of the U.S. Forestry Service, advocates more efficient management and better preservation of forests—which he characterizes as "the art of producing from the forest whatever it can yield for the service of man"—in his book *The Training of a Forester*

1949 Aldo Leopold, professor of wildlife management at the University of Wisconsin, sets forth his influential argument for a land ethic—an environmental ethic based on the inherent value of the natural world and its elements rather than on their economic, recreational, or aesthetic value to humans—in his book *A Sand County Almanac:* "The land ethic," he wrote, "simply enlarges the boundaries of the community to include soils, waters, plants, and animals, or collectively: the land."

1961 A staggering 66,000 whales are killed in this year

1962 Rachel Carson, American naturalist and science writer, brings public attention to the dangers of rampant pesticide use in her influential book, *Silent Spring,* arguing that humans have "allowed these chemicals to be used with little or no advance investigation of their effect on soil, water, wildlife, and man himself."

1963 Carson testifies before Congress, calling for the registration of pesticides and other toxic chemicals and for restrictions on their sale and use

1967 The bald eagle is declared an endangered species in the United States

1970 A new independent environmental oversight agency, the Environmental Protection Agency (EPA), opens its doors

1972 The toxic pesticide DDT is banned for domestic and commercial use in the United States

1973 The Endangered Species Act (ESA) is signed into law with the purpose of protecting vulnerable species from extinction due to "economic growth and development untempered by adequate concern and conservation."

Mountaineer and philosopher Arne Naess introduces the term *deep ecology* to the environmental literature, contrasting it with more shallow mainstream ecological approaches that tend to look for short-term fixes and consumption-driven answers to environmental problems, rather than deep questioning that might lead to the fundamental, long-term changes necessary to preserve the ecological diversity of natural systems

1975 Australian ethicist Peter Singer publishes his influential book *Animal Liberation,* in which he contends that "adult apes, monkeys, dogs, cats, rats, and other animals are more aware of what is happening to them, more self-directing, and, so far as we can tell, at least as sensitive to pain as a human infant."

1980 The U.S. Supreme Court breaks with its tradition of prohibiting the patenting of any products of nature when it rules that a genetically engineered bacterium capable of breaking down crude oil is patentable; Chief Justice Warren Burger writes that "the fact that micro-organisms are alive is without legal significance" and that patent law covers "anything under the sun that is made by man."

1986 The International Whaling Commission (IWC) introduces a moratorium on whale hunting in the hope that threatened species would recover

1988 The U.S. Patent Office grants a patent to the president and fellows of Harvard College for OncoMouse and other "transgenic non-human" mammals

British environmentalist Norman Myers introduces the concept of a hot spot—an area to be singled out for special attention in conservation efforts for two reasons: first, because it contains an exceptionally rich array of plant diversity, and second, because it has already been radically reduced in size by human activities like logging, ranching, and urban and suburban sprawl

1989 The *Exxon Valdez* oil tanker strikes a reef in Prince William Sound, Alaska, setting off the most environmentally devastating oil spill in history

1992 The international Convention on Biological Diversity (CBD) is introduced at the Earth Summit in Rio de Janeiro in June

1993 The Convention on Biological Diversity enters into force in December

1999 Biologists Michael Soulè and John Terborgh call for the creation of land-based mega-reserves in their book *Continental Conservation,* arguing that without them, the "absence of top predators appears to lead inexorably to ecosystem simplification accompanied by a rush of extinctions."

2003 Researchers publishing in the journal *Nature* find that on average, species' ranges are moving toward the poles at some four miles per decade and that spring reproductive events, such as flowering and egg laying, are already shifting 2.3 days earlier per decade due to global climate change; a second research group also reports "a consistent temperature-related shift, or 'fingerprint'" of climate change on a wide range of species, based on a meta-analysis of 143 ecological studies

2005 Conservation International (CI) identifies a total of 34 hot spots that once covered 15.7 percent of the earth's land surface, but only 2.3 percent of it in 2005; these areas contain a staggering percentage of the world's species as endemics (about 50 percent of the world's plant species, 42 percent of terrestrial vertebrate species, and 29 percent of freshwater fish species)

Sister Dorothy Stang, a Catholic nun from Dayton, Ohio, and an outspoken defender of the Amazon rain forest and its native peoples, is shot and killed on a jungle road in the

Brazilian state of Pará after blocking ranchers from taking over a stretch of rain forest that is home to 400 families

2007 The bald eagle is officially removed from the federal list of endangered and threatened species, thanks to the ban on DDT and other protective measures

Xiang Xiang, the first captive giant panda to be released into the wild, is found dead on a forest preserve in China's southwestern Sichuan Province in February, less than a year after his radical relocation

2008 The Svalbard Global Seed Vault—nicknamed the "doomsday vault"—opens its doors to thousands of seed varieties from all over the world; built into a frozen mountain on the remote Norwegian archipelago of Svalbard, this futuristic structure is part of a worldwide effort to protect plant species from annihilation due to climate change, natural disasters, and war

The first attempt to map human impacts on oceans (such as organic pollution, damage from bottom-scraping trawls, and intensive fishing along coral reefs) finds that about 40 percent of ocean areas have been strongly affected by humans, while only 4 percent have been left pristine

The Interior Department of President George W. Bush agrees in May to list the polar bear under the Endangered Species Act but postpones establishing a critical habitat—the area deemed necessary for the survival of the species, and a requirement of the Endangered Species Act

On August 16, federal contractors performing an aerial survey in the Chukchi Sea off the northwest coast of Alaska find nine polar bears swimming in the open sea miles from shore, one of them a full 65 miles (105 km) from land; at least some of the bears are apparently swimming north in a futile attempt to reach the closest stretch of polar ice, which—due to a devastating arctic warming trend—is a full 400 miles (644 km) away

2009 The summer Arctic sea-ice cover is 25 percent less than the average for the period between 1979 and 2000; ice coverage for the years 2007, 2008, and 2009 collectively is the lowest since the satellite record began in 1979

In October, the Obama administration announces a proposal to designate 200,000 square miles (518,000 km²) of Alaskan sea and sea ice as critical habitat for the polar bear

Guyana earns international recognition at the December United Nations Conference on Climate Change in Copenhagen for its cutting-edge plan to protect its rain forest in exchange for international assistance for preservation and sustainable development; the program is held up in support of a broader international program known as REDD (Reduced Emissions from Deforestation and Degradation), which allows industrialized nations to meet CO_2 commitments by funding forest preservation projects in other countries

2010 Genetic material from invasive species of Asian carp is discovered in a harbor in Lake Michigan and in a river within half a mile of the lake; by February, U.S. officials have pledged $78.5 million to fund efforts to keep the carp out of the Great Lakes, which contain 20 percent of the world's freshwater and no natural predators to stop the carp from annihilating native species

Fifty-five countries (including the United States), representing nearly 80 percent of all global CO_2 emissions, have submitted their CO_2 reduction pledges to the REDD program

The Chinese Year of the Tiger sees the six remaining subspecies of the legendary animal on the verge of extinction; tiger populations have been slashed by 97 percent during the last 100 years, with the remaining 3 percent forced onto ever-shrinking and fragmented scraps of habitat

A survey published in the journal *Biological Conservation* finds a decline of nearly 60 percent over the past 35 years in populations of large mammals such as lions, elephants, giraffes, buffalo, and zebras in African national parks; the precipitous drop in numbers is attributed to habitat degradation, insufficient resources and personnel, and the growing bush meat trade

A U.S. federal judge orders that the gray wolf be brought back under federal protection in the states of Montana and Idaho; wolves in those states had been removed from the endangered species list under rules crafted by George W. Bush's administration and upheld by President Obama's Interior Department

Residual DDT in the environment is identified as the most likely reason that populations of the California condor—the nation's largest bird and a longtime resident of the Endangered Species List—are laying eggs with shells so thin that chicks simply cannot survive

On April 20, a violent explosion on the *Deepwater Horizon* drilling rig in the Gulf of Mexico kills 11 workers, injures 17 others, and sets off the largest accidental marine oil spill in history

President Obama announces plans to split the Minerals Management Service into two parts—one in charge of safety and environmental oversight of oil drilling operations and the other responsible for the business end (leasing and revenue collection). Prior to the change, the agency had both reaped the profits from the petroleum industry ($13 billion per year on average) and overseen safety—a system that often allowed the drive for profit to outweigh safety concerns and that permitted petroleum corporations to operate largely unmonitored

International whaling talks collapse in June when Japan refuses to phase out its annual hunt in the Southern Ocean whale sanctuary, where 80 percent of the world's whales go to feed in the summer

Nearly three months after the *Deepwater Horizon* accident, BP Oil is finally successful in capping the plume, but not until almost 5 million barrels—185 million gallons (more than 700 million liters)—of crude oil have spewed into the waters of the Gulf

President Obama signs an executive order in July creating the first federal policy to clean up and protect the oceans, forming a National Ocean Council to coordinate all activities having an impact upon the seas

In August, the Obama administration announces it will overhaul the process by which deepwater drilling permits are granted, requiring more in-depth review of the physical circumstances in each case

On September 19, a full five months after the Deepwater Horizon explosion, the federal government concludes its own pressure tests of the containment cap, deems the cement an effective and permanent seal, and declares the well officially dead

The Interior Department announces in December that it will retract an earlier decision to allow expansion of exploratory drilling into the eastern Gulf and along the Atlantic seaboard, stating that no such operations would be permitted for at least seven years

2011 Congressional leaders and the White House agree to a budget bill rider that would strip the gray wolf of its endangered species status across most of the Northern Rockies

GLOSSARY

anthropocentric human-centered; giving consideration to humans and their interests above all other concerns
archipelago a chain or group of islands clustered close together in an ocean or sea
benzene a colorless, highly flammable liquid hydrocarbon with the simple chemical formula C6H6; benzene is found in crude petroleum and in many petroleum products and is highly toxic and carcinogenic
biocentric viewpoints that expand the scope of ethical consideration to all living things regardless of whether they possess sentience or awareness
biodiversity the number and variety of life-forms living in a particular ecosystem, region, or the entire Earth
biomagnification the process by which a substance (e.g., DDT) is found in the bodies of organisms in higher and higher concentrations as it moves up the food chain
biomass in ecology, biomass refers to the total mass of all living things within a given area or ecosystem
biome a large geographical region characterized by similar climate and vegetation (e.g., a desert or forest) and all living organisms within it
carbon sequestration the natural removal and storage of carbon dioxide from the atmosphere by plants, soil, and water, or any of several geoengineering techniques for removing carbon from the atmosphere in an effort to mitigate the effects of global climate change
cetaceans large aquatic mammals belonging to the order Cetacea, including whales, dolphins, and porpoises
chlorinated hydrocarbon a volatile organic compound containing atoms of carbon, hydrogen, and chlorine; also known as an **organochloride**

Glossary

climate change typically refers to changes in regional and global climate patterns linked to human activities such as the burning of fossil fuels and the clearing of forests and other vegetation

critical habitat the geographic area deemed essential for the survival and recovery of an endangered or threatened species; a required demarcation under the Endangered Species Act

DDT dichlorodiphenyltrichloroethane; a long-lasting synthetic insecticide, banned for domestic and commercial use in the United States in 1972 because of its toxic environmental effects and its risks to human health

deep ecology the belief that ethical consideration should be extended to the entire natural world, including nonliving substances like water, soil, and rock, and that humans and their interests are interdependent with and inseparable from the rest of nature

dichlorodiphenyltrichloroethane see **DDT**

ecosystem a complex collection of relationships among interdependent living and nonliving entities forming a stable system

endemic species that are native or unique to a particular region

extinction the complete destruction or dying off of a particular species, generally considered to occur when the last known individual of the species dies

extirpation local extinction

greenhouse gas a gas that contributes to the greenhouse effect and global climate change by trapping infrared radiation (heat) from the Sun within the Earth's atmosphere

habitat the specific environment in which an organism, species, or community lives or is able to live

hippocampus a seahorse-shaped brain structure in humans and other mammals that plays a critical role in processing emotions and forming long-term memories

hot spot a geographic area of special concern due to a significant level of biodiversity that is threatened with destruction or disruption

hydrocarbon an organic compound containing only carbon and hydrogen atoms

indicator species animals or plants whose presence, relative abundance, or biological composition are reliable measures of particular environmental conditions; often organisms that are especially sensitive to particular environmental changes and can serve as early warning signals for potential environmental problems such as climate change, disease, or pollution

indigenous native to a particular habitat or region

invasive species a nonnative species introduced to an area outside its natural range, often with harmful consequences for native species

keystone species an animal or plant species whose importance to the health of its ecosystem is much greater than would be expected based on a simple analysis of its relative numbers or total biomass

land ethic an ethical framework, set forth by Aldo Leopold in his book *A Sand County Almanac,* that bases conservation decisions on the inherent value of the natural world and its elements rather than on their economic, recreational, or aesthetic value to humans

macroecology the subfield of ecology that seeks to understand patterns of diversity and abundance across large geographic scales

mass extinction event a period characterized by a precipitous decrease in the diversity and abundance of life

mega-reserve a protected area of land or water large enough to shelter entire ecosystems in a relatively undisturbed state

monoculture the practice of cultivating a single crop on a piece of land, thereby discouraging diversity; a stretch of land dominated by a single plant species

moratorium a temporary ban or suspension of a specific activity

mutualistic a relationship between two organisms or species that is beneficial to both

Glossary

mysticetes also known as the great whales, these toothless, filter-feeding whales (e.g., blue and humpback whales) strain water through baleen to capture small prey

niche the particular role, function, or place held by an organism within its community or ecosystem

odontocetes the toothed whales, including sperm whales, orca, and dolphins

organochloride a volatile organic compound containing atoms of carbon, hydrogen, and chlorine; also known as a **chlorinated hydrocarbon**

overexploitation when a particular plant or animal is harvested or hunted by humans to such a degree that their overall population is threatened or extinguished

phytoplankton tiny, free-floating aquatic plants that are a basic food source in many marine ecosystems

poaching illegal hunting, fishing, or trapping

rain forest a forest characterized by heavy annual precipitation; rain forests are home to the planet's richest stores of plant and animal diversity and absorb large amounts of carbon dioxide from the atmosphere, thus slowing the rate of global climate change

sentient the quality of possessing sense organs and a nervous system capable of feeling or perceiving sensory input

species a group of organisms that share certain characteristics and can interbreed

speciesism a term used to describe prejudice against non-human animals, similar to racism or sexism

sustainable development economic, agricultural, and industrial development that maintains and protects the natural environment and its capacity to restore itself

terrestrial land (as opposed to aquatic) environments; species that live primarily on land

umbrella species species selected for special conservation efforts because their land or resource requirements are large enough that protecting them will automatically protect other species living within their habitat

understory vegetation growing in the shade of a forest canopy

utilitarianism the ethical theory that seeks the greatest benefit for the greatest number

vascular plant a plant with internal vessels for transporting nutrients and water

wetland a swamp, marsh, or other area where the soil is saturated by water much or all of the year; these critically important ecosystems shelter a rich variety of plants and wildlife, while also preventing floods, filtering water, and protecting land from erosion

FURTHER RESOURCES

Chapter 1: Conserving Biodiversity
The story of the Svalbard Global Seed Vault comes primarily from press reports and from information provided by the Global Crop Diversity Trust. Perspectives on the relationship between humans and the rest of the natural world are drawn from original philosophical works such as Aristotle's *Politics*, Peter Singer's *Animal Liberation*, and Aldo Leopold's *A Sand County Almanac*. The IUCN Red List is an invaluable resource for data on threatened and endangered species. All sources are detailed below.

Aristotle. *Politics: A Treatise on Government.* Translated by William Ellis. Self-published, 2010.
Crichton, Michael. "Patenting Life." *New York Times* (13 February 2007). Available online. URL: http://www.nytimes.com/2007/02/13/opinion/13crichton.html?ref=michael_crichton. Accessed March 27, 2011.
DePalma, Anthony. "Texcoco Journal; The 'Slippery Slope' of Patenting Farmers' Crops." *New York Times* (24 May 2000). Available online. URL: http://www.nytimes.com/2000/05/24/world/texcoco-journal-the-slippery-slope-of-patenting-farmers-crops.html. Accessed March 27, 2011.
Drengson, Alan. "Deep Ecology Movement." Available online. URL: http://www.deepecology.org/movement.htm. Accessed March 27, 2011.
Fowler, Cary. "The Svalbard Global Seed Vault: Securing the Future of Agriculture" (26 February 2008). Available online. URL: http://www.croptrust.org/documents/Svalbard%20opening/New%20EMBARGOED-Global%20Crop%20Diversity%20Trust%20Svalbard%20Paper.pdf. Accessed on March 27, 2011.
Global Crop Diversity Trust. "Svalbard Global Seed Vault." Available online. URL: http://www.croptrust.org/main/arcticseedvault.php?itemid=211. Accessed March 27, 2011.

Leopold, Aldo. *A Sand County Almanac*. 2d ed. New York: Oxford University Press, 1968.

Martin, Douglas. "Bent Skovmand, Seed Protector, Dies at 61." *New York Times* (14 February 2007). Available online. URL: http://www.nytimes.com/2007/02/14/science/14skovmand.html?scp=1&sq=Bent+Skovmand%2C+Seed+Protector% 2C+Dies+at+61&st=nyt. Accessed March 27, 2011.

Pinchot, Gifford. *The Training of a Forester*. New York: J. B. Lippincott, 1937.

Pollack, Andrew. "U.S. Says Genes Should Not Be Eligible for Patents." *New York Times* (29 October 2010). Available online. URL: http://www.nytimes.com/2010/10/30/business/30drug.html?ref=genetic engineering. Accessed March 27, 2011.

Regan, Tom. *The Case for Animal Rights: Updated with a New Preface*. Berkeley: University of California Press, 2004.

Revkin, Andrew C. "Global Effort to Save Endangered Crops Gets $37.5 Million Infusion." *New York Times* (19 April 2007). Available online. URL: http://www.nytimes.com/2007/04/19/science/earth/19farm.ready.html? scp=1&sq=Global+Effort+to+Save+Endangered+Crops+Gets+%2437.5+Million +Infusion.%94+&st=nyt. Accessed March 27, 2011.

Rosenthal, Elisabeth. "Near Arctic, Seed Vault Is a Fort Knox of Food." *New York Times* (29 February 2008). Available online. URL: http://www.nytimes.com/2008/02/29/world/europe/29seeds.html?_r=1&scp=1&sq=Near%20Arctic,%20Seed%20Vault%20Is%20a%20Fort%20Knox%20of%20Food.%E2%80%9D%20&st=cse. Accessed March 27, 2011.

Rowan, Andrew N. "Formulation of Ethical Standards for Use of Animals in Medical Research." *Toxicology Letters* 68, nos. 1–2 (May 1993): 63–71.

Shorett, Peter. "Of Transgenic Mice and Men." *GeneWatch* 15, no. 5 (September 2002). Available online. URL: http://www.councilforresponsiblegenetics.org/ViewPage.aspx?pageId=167. Accessed March 27, 2011.

Singer, Peter. *Animal Liberation*. 2d ed. New York: New York Review/Random House, 1990.

Vié, Jean-Christophe, Craig Hilton-Taylor, and Simon N. Stuart, eds. *Wildlife in a Changing World: An Analysis of the 2008 IUCN Red List of Threatened Species*. Gland, Switzerland: IUCN, 2009. Available online. URL: http://data.iucn.org/dbtw-wpd/edocs/RL-2009-001.pdf. Accessed March 27, 2011.

Chapter 2: Habitat Destruction and Restoration

Recent media coverage provides up-to-date information on the precarious state of the polar bear, while data on shrinking Arctic sea ice comes largely from the National Oceanic and Atmospheric Administration (NOAA). Information on deforestation in the Amazon is drawn largely from original research published in the journals *Science* and *Proceedings of the National Academy of Sciences* (*PNAS*), and Sister Dorothy Stang's life and death are chronicled in a series of stories in the *New York Times*. All sources are detailed below.

"As Arctic Sea Ice Melts, Experts Expect New Low." Associated Press (27 August 2008). Available online. URL: http://www.nytimes.com/2008/08/28/science/earth/28seaice.html. Accessed March 27, 2011.

Barcott, Bruce. "Twilight of the Ice Bear." *New York Times* (10 December 2009). Available online. URL: http://www.nytimes.com/2009/12/13/books/review/Barcott-t.html?scp=1&sq=%22twilight%20of%20the%20ice%20bear%22&st=cse. Accessed March 27, 2011.

Barringer, Felicity. "Polar Bear Is Made a Protected Species." *New York Times* (15 May 2008). Available online. URL: http://www.nytimes.com/2008/05/15/us/15polar.html?scp=1&sq=Polar%20Bear%20Is%20Made%20a%20Protected%20Species&st=cse. Accessed March 27, 2011.

Barrionuevo, Alexei. "Forest Plan in Brazil Bears the Traces of an Activist's Vision." *New York Times* (21 December 2008). Available online. URL: http://www.nytimes.com/2008/12/22/world/americas/22brazil.html. Accessed March 27, 2011.

"Breathing Room for the Bear." *New York Times* editorial (23 October 2009). Available online. URL: http://www.nytimes.com/2009/10/24/opinion/24sat4.html?scp=1&sq=Breathing%20Room%20for%20the%20Bear&st=cse. Accessed March 27, 2011.

Broder, John M. "Polar Bear Habitat Proposed for Alaska." *New York Times* (22 October 2009). Available online. URL: http://www.nytimes.com/2009/10/23/science/earth/23bear.html?scp=1&sq=Polar%20Bear%20Habitat%20Proposed%20for%20Alaska.&st=cse. Accessed March 27, 2011.

Conservation International. "Hotspots Defined." Available online. URL: http://www.biodiversityhotspots.org/xp/Hotspots/hotspotsScience/pages/ hotspots_defined.aspx. Accessed March 27, 2011.

Feeley, Kenneth J., and Miles R. Silman. "Extinction Risks of Amazonian Plant Species." *PNAS* 106, no. 30 (28 July 2009):

12,382–12,387. Available online. URL: http://www.pnas.org/content/106/30/12382.full. Accessed March 27, 2011.

Friedman, Thomas L. "Trucks, Trains and Trees." *New York Times* (11 November 2009). Available online. URL: http://www.nytimes.com/2009/11/11/opinion/11friedman.html. Accessed March 27, 2011.

Gies, Erica. "Guyana Offers a Model to Save Rain Forest." *New York Times* (8 December 2009). Available online. URL: http://www.nytimes.com/2009/12/08/business/global/08iht-rbogeco.html?fta=y. Accessed March 27, 2011.

———. "U.K.-Based Financier Invests in Guyana's Rain Forest." *New York Times* (8 December 2009). Available online. URL: http://www.nytimes.com/2009/12/08/business/global/08iht-rbogguy1.html?fta=y. Accessed March 27, 2011.

Hubbell, Stephen P., et al. "How Many Tree Species Are There in the Amazon and How Many of Them Will Go Extinct?" *PNAS* 105, suppl. 1 (12 August 2008): 11,498–11,504. Available online. URL: http://www.pnas.org/content/105/suppl.1/11498.full. Accessed March 27, 2011.

Jha, Shalene, and Christopher W. Dick. "Shade Coffee Farms Promote Genetic Diversity of Native Trees." *Current Biology* 18, no. 24 (23 December 2008): R1,126–R1,128.

Myers, Norman, et al. "Biodiversity Hotspots for Conservation Priorities." *Nature* 403 (24 February 2000): 853–858. Available online. URL: http://people.biology.ufl.edu/osenberg/courses/Seminar_in_Ecology/ 2005_Fall/readings/Myers%20et%20al%202000%20(Nature).pdf. Accessed March 27, 2011.

National Oceanic and Atmospheric Administration. "Future of Arctic Sea Ice and Global Impacts." Available online. URL: http://www.arctic.noaa.gov/future/index.html. Accessed March 27, 2011.

———. "Arctic Free of Summer Sea Ice within 30 Years?" Available online. URL: http://www.arctic.noaa.gov/future/sea_ice.html. Accessed March 27, 2011.

Nepstad, Daniel, et al. "The End of Deforestation in the Brazilian Amazon." *Science* 326, no. 5,958 (4 December 2009): 1,350–1,351. Available online. URL: http://www.hks.harvard.edu/var/ezp_site/storage/fckeditor/file/pdfs/centers-programs/centers/cid/ssp/docs/events/workshops/2010/foodsecurity/Neptstad_Science_2009.pdf. Accessed March 27, 2011.

"Polar Bears Found Swimming Miles from Alaskan Coast." *ScienceDaily* (26 August 2008). Available online. URL: http://www.science

daily.com/releases/2008/08/080825210415.htm. Accessed March 27, 2011.

Revkin, Andrew C. "Nations Near Arctic Declare Polar Bears Threatened by Climate Change." *New York Times* (19 March 2009). Available online. URL: http://www.nytimes.com/2009/03/20/science/earth/20bears.html?scp= 1&sq=Nations%20Near%20Arctic%20Declare%20Polar%20Bears%20Threatened%20by%20Climate%20Change&st=cse. Accessed March 27, 2011.

———. "U.S. Curbs Use of Species Act in Protecting Polar Bear." *New York Times* (9 May 2009). Available online. URL: http://www.nytimes.com/2009/05/09/science/earth/09bear.html?scp=1&sq=U.S.%20Curbs%20Use%20of%20Species%20Act%20in%20Protecting%20Polar%20Bear&st=cse. Accessed March 27, 2011.

Rohter, Larry. "Brazil, Bowing to Protests, Reopens Logging in Amazon." *New York Times* (13 February 2005). Available online. URL: http://www.nytimes.com/2005/02/13/international/americas/13amazon.html. Accessed March 27, 2011.

———. "Struggling to Save His Amazon, from the Top of a Death List." *New York Times* (30 December 2006). Available online. URL: http://www.nytimes.com/2006/12/30/world/americas/30feitosa.html. Accessed March 27, 2011.

Rosenthal, Elisabeth. "Climate Talks Near Deal to Save Forests." *New York Times* (15 December 2009). Available online. URL: http://www.nytimes.com/2009/12/16/science/earth/16forest.html?scp=8&sq=biodiversity&st=cse. Accessed March 27, 2011.

"Sister Dorothy Stang." *New York Times* (3 May 2010). Available online. URL: http://topics.nytimes.com/top/reference/timestopics/people/s/sister_dorothy_stang/index.html?offset=0&s=newest. Accessed March 27, 2011.

UN Food and Agriculture Organization. "Livestock's Long Shadow: Environmental Issues and Options." 2006. Available online. URL: ftp://ftp.fao.org/docrep/fao/010/a0701e/a0701e00.pdf. Accessed March 27, 2011.

University of Illinois at Chicago Office of Public Affairs. "Birds and Bats Sow Tropical Seeds." (13 February 2005). Available online. URL: http://tigger.uic.edu/htbin/cgiwrap/bin/newsbureau/cgi-bin/index.cgi?from=Releases&to=Release&id=1215 &start=1119958944&end=1127734944&topic=0&dept=0. Accessed March 27, 2011.

University of Michigan News Service. "Shade Coffee Benefits More Than Birds." (22 December 2008). Available online. URL: http://

ns.umich.edu/htdocs/releases/story.php?id=6908. Accessed March 27, 2011.

USAID. "Conserving Biodiversity in the Amazon Basin." (May 2005). Available online. URL: http://pdf.usaid.gov/pdf_docs/PNADF441.pdf. Accessed March 27, 2011.

World Bank. "Assessment of the Risk of Amazon Dieback." (4 February 2010). Available online. URL: http://docs.google.com/viewer?a=v&q=cache:VPPTNEi6KpwJ:www.bicusa.org/en/Document.101982.aspx+world+bank+assessment+of+the+risk+of+amazon+dieback&hl=en&gl=us&pid=bl&srcid=ADGEESg6T1s2x_X0XXkuw3aRks0qlJGQjWh3aD6RPc74J0VnMlJeHUDDuq1efs1aIY0hjSinp5F8-jS48UlQrUYs8pBo5iJrnOeh9VRN2ffJZO9VECKMACxIQ0-m0zazj3u unwQ2q0oI&sig=AHIEtbTAQc1F7k6edCLvv63BQDmR2iE77Q. Accessed March 27, 2011.

World Wildlife Fund. "Polar Bears and Sea Ice Loss: Questions and Answers." (15 September 2008). Available online. URL: http://www.worldwildlife.org/species/finder/polarbear/WWFBinaryitem10113.pdf. Accessed March 27, 2011.

Chapter 3: Toxic Contamination and Cleanup

Information for this chapter is drawn primarily from press reports and journal articles about the BP *Deepwater Horizon* disaster and its environmental impact; historical and policy documents on the DDT ban and the creation of the EPA; and Rachel Carson's influential volume on the potential health and environmental effects of toxic pollution, *Silent Spring*. All sources are detailed below.

American Eagle Foundation. "Bald Eagle: The U.S.A.'s National Symbol." Available online. URL: http://web.archive.org/web/20071206030939/http://www.eagles.org/moreabout.html. Accessed March 27, 2011.

Broder, John M. "Blunders Abounded Before Gulf Spill, Panel Says." *New York Times* (5 January 2011). Available online. URL: http://www.nytimes.com/2011/01/06/science/earth/06spill.html?ref=gulfofmexico2010. Accessed March 27, 2011.

———. "Investigator Finds No Evidence That BP Took Shortcuts to Save Money." *New York Times* (8 November 2010). Available online. URL: http://www.nytimes.com/2010/11/09/us/09spill.html?scp=1&sq=bartlit&st=cse. Accessed March 27, 2011.

———. "Panel Says Firms Knew of Cement Flaws Before Spill." *New York Times* (28 October 2010). Available online. URL: http://www.

nytimes.com/2010/10/29/us/29spill.html?ref=gulf_of_mexico _2010. Accessed March 27, 2011.

———. "Report Faults BP and Contractors for Rig Explosion and Spill." *New York Times* (17 November 2010). Available online. URL: http://www.nytimes.com/2010/11/18/us/18BP.html?scp=3&sq=bp spill&st=cse. Accessed March 27, 2011.

Camilli, Richard, et al. "Tracking Hydrocarbon Plume Transport and Biodegradation at Deepwater Horizon." *Science* 330, no. 6,001 (19 August 2010): 201–204. Available online. URL: http://www.reefrelief founders.com/drilling/wp-content/uploads/2010/08/Woods-Hole-Tracking-Hydrocarbon-plumes-paper.pdf. Accessed March 27, 2011.

Carson, Rachel. *Silent Spring: With an Introduction by Vice President Al Gore.* New York: Houghton Mifflin, 1994.

Gillis, Justin, and John Collins Rudolf. "Oil Plume Is Not Breaking Down Fast, Study Says." *New York Times* (19 August 2010). Available online. URL: http://www.nytimes.com/2010/08/20/science/earth/20plume.html?ref=gulf_of_mexico_2010. Accessed March 27, 2011.

"Gulf of Mexico Oil Spill (2010)." *New York Times* (24 November 2010). Available online. URL: http://topics.nytimes.com/top/reference/timestopics/subjects/o/oil_spills/gulf_of_mexico_2010/index.html?scp=2&sq=BPspill&st=cse. Accessed March 27, 2011.

Hynes, H. Patricia. *The Recurring Silent Spring.* Oxford, U.K.: Pergamon Press, 1989.

Kerr, Richard A. "Report Paints New Picture of Gulf Oil." *Science Now* (19 August 2010). Available online. URL: http://news.sciencemag.org/sciencenow/2010/08/report-paints-new-picture-of-gul.html. Accessed March 27, 2011.

Lear, Linda. "Rachel Carson's Biography." *The Life and Legacy of Rachel Carson.* Available online. URL: http://www.rachelcarson.org/. Accessed March 27, 2011.

Lewis, Jack. "The Birth of EPA." *EPA Journal* (November 1985). Available online. URL: http://www.epa.gov/history/topics/epa/15c.htm. Accessed March 27, 2011.

Lubchenco, Jane, et al. "BP Deepwater Horizon Oil Budget: What Happened to the Oil?" (4 August 2010). Available online. URL: http://documents.nytimes.com/noaa-usgs-report-shows-gulf-of-mexico-oil-spill-poses-little-additional-risk#document/p1. Accessed March 27, 2011.

Matthiessen, Peter. "Environmentalist Rachel Carson." *Time* (29 December 1999). Available online. URL: http://www.time.com/time/printout/0,8816,990622,00.html. Accessed March 27, 2011.

McLaughlin, Dorothy. "Silent Spring Revisited." *Frontline,* PBS/WGBH. Available online. URL: http://www.pbs.org/wgbh/pages/frontline/shows/nature/disrupt/sspring.html. Accessed March 27, 2011.

Moir, John. "New Hurdle for California Condors May Be DDT from Years Ago." *New York Times* (15 November 2010). Available online. URL: http://www.nytimes.com/2010/11/16/science/16condors.html?ref=science. Accessed March 27, 2011.

Rudolf, John Collins. "Scientists Back Early Government Report on Gulf Spill." *New York Times* (23 November 2010). Available online. URL: http://www.nytimes.com/2010/11/24/science/earth/24spill.html?ref=gulfofmexico2010. Accessed March 27, 2011.

Schwartz, John. "U.S. Sues Companies for Spill Damages." *New York Times* (15 December 2010). Available online. URL: http://www.nytimes.com/2010/12/16/us/16suit.html?ref=gulfofmexico2010. Accessed March 27, 2011.

U.S. Environmental Protection Agency. "DDT Ban Takes Effect." (31 December 1972). Available online. URL: http://www.epa.gov/history/topics/ddt/01.htm. Accessed March 27, 2011.

———. "Palos Verdes Shelf." (5 October 2010). Available online. URL: http://yosemite.epa.gov/r9/sfund/r9sfdocw.nsf/7508188dd3c99a2a8825742600743735/e61d5255780dd68288257007005e9422!OpenDocument#threats. Accessed March 27, 2011.

U.S. Fish and Wildlife Service. "Contaminants and Birds." Available online. URL: http://www.fws.gov/contaminants/examples/AlaskaPeregrine.cfm. Accessed March 27, 2011.

Chapter 4: Species Overexploitation, Protection, and Captive Breeding

Key sources of information on endangered species include the World Wildlife Fund and original research published in the journal *BioScience*. Accounts of recent political controversies over the hunting of whales, tigers, and gray wolves are drawn from a range of relevant press reports, while information on captive breeding comes largely from the Smithsonian Conservation Biology Institute and research published in *Science*. All sources are detailed below.

Angier, Natalie. "Save a Whale, Save a Soul, Goes the Cry." *New York Times* (26 June 2010). Available online. URL: http://www.nytimes.com/2010/06/27/weekinreview/27angier.html?scp=1&sq=Save%20a%20Whale,%20Save%20a%20Soul,%20Goes%20the%20Cry&st=cse. Accessed March 27, 2011.

Further Resources

Araki, Hitoshi, Becky Cooper, and Michael S. Blouin. "Genetic Effects of Captive Breeding Cause a Rapid, Cummulative Fitness Decline in the Wild." *Science* 318, no. 5,847 (5 October 2007): 100–103. Available online. URL: http://www.cnr.uidaho.edu/fish415/Wilhelm%20files/Araki%20et%20al%20trout%20reproduction.pdf. Accessed March 27, 2011.

Barringer, Felicity. "Judge Orders Protection for Wolves in 2 States." *New York Times* (5 August 2010). Available online. URL: http://www.nytimes.com/2010/08/06/science/earth/06wolf.html?scp=1&sq=like%20saying%20an%20orange%20is%20an%20orange%20only%20when%20it%20is%20hanging%20on%20a%20tree&st=cse. Accessed March 27, 2011.

Beament, Emily. "Nations Fail to Reach Commercial Whaling Agreement." *The Independent* (23 June 2010). Available online. URL: http://www.independent.co.uk/environment/nature/nations-fail-to-reach-commercial-whaling-agreement-2008314.html. Accessed March 27, 2011.

"Captive-Bred Panda Dies in the Wild." *New York Times* (1 June 2007). Available online. URL: http://query.nytimes.com/gst/fullpage.html?res=9F05E5D71530F932A35755C0A9619C8B63&scp=1&sq=Captive-Bred%20Panda%20Dies%20in%20the%20Wild.%E2%80%9D%20&st=cse. Accessed March 27, 2011.

"Captive Breeding Success Stories." *Nature,* PBS. Available online. URL: http://www.pbs.org/wnet/nature/episodes/the-loneliest-animals/captive-breeding-success-stories/ 4920/. Accessed March 27, 2011.

"China to Build Training Center for Pandas to Survive in Wild." Associated Press (20 May 2010). Available online. URL: http://www.usatoday.com/tech/news/2010-05-20-pandas-wild_N.htm. Accessed March 27, 2011.

Dean, Cornelia. "The Fall and Rise of the Right Whale." *New York Times* (16 March 2009). Available online. URL: http://www.nytimes.com/2009/03/17/science/17whal.html?_r=1&scp=1&sq=The%20Fall%20and%20Rise%20of%20the%20Right%20Whale&st=cse. Accessed March 27, 2011.

Dinerstein, Eric, et al. "The Fate of Wild Tigers." *BioScience* 57, no. 6 (June 2007): 508–514. Available online. URL: http://legacy.ucpress.net/doi/full/10.1641/B570608. Accessed March 27, 2011.

Elegant, Simon. "When Pandas Go Wild." *Time* (6 January 2007). Available online. URL: http://www.time.com/time/health/article/0,8599,1574819,00.html. Accessed March 27, 2011.

Jackson, Patrick. "Tigers and Other Farmyard Animals." BBC News (29 January 2010). Available online. URL: http://news.bbc.co.uk/2/hi/asia-pacific/8487122.stm. Accessed March 27, 2011.

Jolly, David. "Under Pressure, Commission Discusses Lifting Whaling Ban." *New York Times* (21 June 2010). Available online. URL: http://www.nytimes.com/2010/06/22/world/22whale.html?scp=1&sq=Under%20Pressure,%20Commission%20Discusses%20Lifting%20Whaling%20Ban.%E2%80%9D%20&st=cse. Accessed March 27, 2011.

———. "Whaling Talks in Morocco Fail to Produce Reductions." *New York Times* (23 June 2010). Available online. URL: http://www.nytimes.com/2010/06/24/world/24whale.html?scp=1&sq=Whaling%20Talks%20in%20Morocco%20Fail%20to%20Produce%20Reductions&st=cse. Accessed March 27, 2011.

Macartney, Jane. "Harsh Life in the Wild Kills Panda Bred in Captivity." *Times* (London, 1 June 2007). Available online. URL: http://www.timesonline.co.uk/tol/news/world/asia/article1867822.ece. Accessed March 27, 2011.

Marsh, Bill. "Fretting About the Last of the World's Biggest Cats." *New York Times* (6 March 2010). Available online. URL: http://www.nytimes.com/2010/03/07/weekinreview/07marsh.html. Accessed March 27, 2011.

Pear, Robert. "With a Spending Deal in Hand, Lawmakers Now Turn to the Details." *New York Times,* 10 April 2011. Available online. URL: http://www.nytimes.com/2011/04/11/us/11budget.html?scp=1&sq=%22gray%20wolf%22&st=cse. Accessed April 11, 2011.

Robbins, Jim. "For Wolves, a Recovery May Not Be the Blessing It Seems." *New York Times* (6 February 2007). Available online. URL: http://www.nytimes.com/2007/02/06/science/06wolf.html?scp=1&sq=For%20Wolves,%20a%20Recovery%20May%20Not%20Be%20the%20Blessing%20It%20Seems&st=cse. Accessed March 27, 2011.

———. "Gray Wolf Will Lose Protection in Part of U.S." *New York Times* (6 March 2009). Available online. URL: http://www.nytimes.com/2009/03/07/science/earth/07wolves.html?scp=1&sq=The+recovery+of+the+gray+wolf+throughout+significant+portions+of+its+historic+range+is+one+of+the+great+success+stories+of+the+Endangered+Species+Act&st=nyt. Accessed March 27, 2011.

Smithsonian Conservation Biology Institute. "Endangered Species Science: Captive Breeding. Available online. URL: http://nationalzoo.si.edu/scbi/EndangeredSpecies/CapBreedPops/default.cfm. Accessed March 27, 2011.

Taylor, Phil. "Wolf Delisting Survives Budget Fight, as Settlement Crumbles." *New York Times,* 11 April 2011. Available online. URL: http://www.nytimes.com/gwire/2011/04/11/11greenwire-wolf-delisting-survives-budget-fight-as-settle-61474.html?scp=3&sq=%22gray%20wolf%22&st=cse. Accessed April 11, 2011.

"Tigers in Serious Trouble Around the World, Including U.S." *ScienceDaily* (11 February 2010). Available online. URL: http://www.sciencedaily.com/releases/2010/02/100210124813.htm. Accessed March 27, 2011.

"Viable Tiger Populations, Tiger Trade Incompatible." *ScienceDaily* (7 June 2007). Available online. URL: http://www.sciencedaily.com/releases/2007/06/070605120929.htm. Accessed March 27, 2011.

"Victory for Wolves." *New York Times* editorial (6 August 2010). Available online. URL: http://www.nytimes.com/2010/08/07/opinion/07sat4.html?ref=wolves. Accessed March 27, 2011.

"Whaling: The Japanese Position." BBC News (15 January 2008). Available online. URL: http://news.bbc.co.uk/2/hi/asia-pacific/7153594.stm. Accessed March 27, 2011.

World Wildlife Fund. "Save the Whale, Save the Southern Ocean: Why Preventing Whaling in the Southern Ocean Is Crucial for the World's Whales." (June 2010). Available online. URL: http://assets.panda.org/downloads/wwf_savethewhale_web.pdf. Accessed March 27, 2011.

———. "Tiger Overview." Available online. URL: http://www.worldwildlife.org/species/finder/tigers/index.html. Accessed March 27, 2011.

———. "Whales and Dolphins: Old Dangers Persist, New Ones Have Appeared." Available online. URL: http://www.worldwildlife.org/species/finder/cetaceans/whalesanddolphins.html. Accessed March 27, 2011.

———. "Year of the Tiger Begins with Big Cats in Serious Trouble around the World, Including Here in the U.S." (10 February 2010). Available online. URL: http://www.worldwildlife.org/who/media/press/2010/WWFPresitem15288.html. Accessed March 27, 2011.

Chapter 5: Invasive Species and Their Impact on Diversity

Overviews from the U.S. Department of Agriculture and the U.S. Department of the Interior provide historical background on the invasion of Guam by the brown tree snake, while the encroachment of Asian carp on the Great Lakes watershed is detailed in a series of articles in the *New York Times*. An alternative perspective on nonnative species is presented in the journal *PNAS*, and the effects of global climate change on native and nonnative species are investigated in two studies published in the journal *Nature*. All sources are detailed below.

Angelos, James. "Woodsman, Don't Spare That Tree." *New York Times* (11 May 2008). Available online. URL: http://www.nytimes.com/2008/05/11/nyregion/thecity/11pres.html. Accessed March 27, 2011.

Barry, Dan. "On an Infested River, Battling Invaders Eye to Eye." *New York Times* (14 September 2008). Available online. URL: http://www.nytimes.com/2008/09/15/us/15land.html?ref=invasive_species. Accessed March 27, 2011.

———. "Weed Heroes: The War on the Invader Cogongrass." *New York Times* (20 September 2009). Available online. URL: http://www.nytimes.com/2009/09/21/us/21land.html?ref=invasive_species. Accessed March 27, 2011.

Christopher, Tom. "Can Weeds Help Solve the Climate Crisis?" *New York Times* (29 June 2008). Available online. URL: http://www.nytimes.com/2008/06/29/magazine/29weeds-t.html?ref=invasive_species. Accessed March 27, 2011.

"Closing the Carp Highway." *New York Times* editorial (18 February 2010). Available online. URL: http://www.nytimes.com/2010/02/18/opinion/18thur4.html?ref=invasive_species. Accessed March 27, 2011.

Davey, Monica. "Be Careful What You Fish For." *New York Times* (12 December 2009). Available online. URL: http://www.nytimes.com/2009/12/13/weekinreview/13davey.html?ref=invasive_species. Accessed March 27, 2011.

Lendon, Brad. "Tylenol-Loaded Mice Dropped from Air to Control Snakes." CNN.com (7 September 2010). Available online. URL: http://news.blogs.cnn.com/2010/09/07/tylenol-loaded-mice-dropped-from-air-to-control-snakes/. Accessed March 27, 2011.

———. "U.S. Officials Plan $78.5 Million Effort to Keep Dangerous Carp Out of Great Lakes." *New York Times* (9 February 2010). Available online. URL: http://www.nytimes.com/2010/02/09/science/09asiancarp.html?ref=invasive_species. Accessed March 27, 2011.

Olsen, Erik. "Vacuuming the Reef." *New York Times* (2008). Available online. URL: http://video.nytimes.com/video/2009/02/19/science/1194837960943/ vacuuming-the-reef.html?ref=invasive_species. Accessed March 27, 2011.

Parmesan, Camille, and Gary Yohe. "A Globally Coherent Fingerprint of Climate Change Impacts Across Natural Systems." *Nature* 421 (2 January 2003): 37–42. Available online. URL: http://www.arp.harvard.edu/sci/climate/journalclub/Parmesan.pdf. Accessed March 27, 2011.

Podger, Pamela J. "Got Weeds? These Sheep Will Make House Calls." *New York Times* (26 October 2008). Available online. URL: http://www.nytimes.com/2008/10/27/us/27weeds.html?ref=invasive_species. Accessed March 27, 2011.

Revkin, Andrew C. "Warming Is Found to Disrupt Species." *New York Times* (2 January 2003). Available online. URL: http://query.nytimes.com/gst/fullpage.html?res=940DE2DB103FF931A35752C0A9659C8B63&ref=invasive_species. Accessed March 27, 2011.

Robbins, Jim. "In a Warmer Yellowstone Park, a Shifting Environmental Balance." *New York Times* (18 March 2008). Available online. URL: http://www.nytimes.com/2008/03/18/science/18griz.html?ref=invasive_species. Accessed March 27, 2011.

Root, Terry L., et al. "Fingerprints of Global Warming on Wild Animals and Plants." *Nature* 421 (2 January 2003): 57–60.

Saulny, Susan. "Carp DNA Is Found in Lake Michigan." *New York Times* (20 January 2010). Available online. URL: http://query.nytimes.com/gst/fullpage.html?res=9401EED9123BF933A15752C0A9669D8B63&ref=invasive_species. Accessed March 27, 2011.

Sax, Dov F., and Steven D. Gaines. "Species Invasions and Extinction: The Future of Native Biodiversity on Islands." *PNAS* 105, suppl. 1 (12 August 2008): 11,490–11,497. Available online. URL: http://www.pnas.org/content/105/suppl.1/11490.full. Accessed March 27, 2011.

Svoboda, Elizabeth. "The Unintended Consequences of Changing Nature's Balance." *New York Times* (16 February 2009). Available online. URL: http://www.nytimes.com/2009/02/17/science/17isla.html?ref=invasive_species. Accessed March 27, 2011.

University of Washington News and Information. "Brown Tree Snake Could Mean Guam Will Lose More Than Its Birds." (7 August 2008). Available online. URL: http://uwnews.org/article.asp?articleid=43191. Accessed March 27, 2011.

U.S. Department of Agriculture. Animal and Plant Health Inspection Service. "The Brown Tree Snake." (November 2001). Available online. URL: http://www.aphis.usda.gov/lpa/pubs/fsheet_faq_notice/fs_wsbtsnake.html. Accessed March 27, 2011.

U.S. Department of the Interior. U.S. Geological Survey. "The Brown Tree Snake on Guam." (2001). Available online. URL: http://www.fort.usgs.gov/Resources/Education/BTS/. Accessed March 27, 2011.

Zimmer, Carl. "Friendly Invaders." *New York Times* (8 September 2008). Available online. URL: http://www.nytimes.com/2008/09/09/science/09inva.html?_r=1&ref=invasive_species. Accessed March 27, 2011.

Chapter 6: Ecosystem Disruption and the Call for Mega-reserves

Articles in *Science, ScienceDaily,* and *Scientific American* provide rich background on protective relationships among elephants, ants, and acacia on the African savanna; a series of articles in the *New York Times* explores a range of issues with no-take marine reserves and complex interactions among organisms in reef ecosystems. Arguments for land-based mega-reserves come primarily from studies in *Conservation Biology* and *TRENDS in Ecology and Evolution,* while the pervasive human impact on the world's oceans is revealed in a pioneering article in the journal *Science* (Halpern, et al). All sources are detailed below.

"Africa's Biggest Mammals Key to Ant-Plant Teamwork." *ScienceDaily* (16 January 2008). Available online. URL: http://www.sciencedaily.com/releases/2008/01/080110144845.htm. Accessed March 27, 2011.

"Africa's National Parks Hit by Mammal Declines." *ScienceDaily* (13 July 2010). Available online. URL: http://www.sciencedaily.com/releases/2010/07/100712141851.htm. Accessed March 27, 2011.

Biello, David. "Of Ants, Elephants, and Acacias: A Tale of Ironic Interdependence." *Scientific American* (10 January 2008). Available online. URL: http://www.scientificamerican.com/article.cfm?id=of-ants-elephants-and-acacias. Accessed March 27, 2011.

Dean, Cornelia. "Coral Reefs and What Ruins Them." *New York Times* (26 February 2008). Available online. URL: http://www.nytimes.com/2008/02/26/science/earth/26reef.html. Accessed March 27, 2011.

———. "In Life's Web, Aiding Trees Can Kill Them." *New York Times* (11 January 2008). Available online. URL: http://www.nytimes.com/2008/01/11/science/11ants.html?scp=3&sq=elephants and acacia&st=cse. Accessed March 27, 2011.

Fountain, Henry. "Coral Trout Thrive in Protected Parts of Reef." *New York Times* (1 July 2008). Available online. URL: http://www.nytimes.

com/2008/07/01/science/01obreef.html?scp=1&sq=%22coral%20 trout%20thrive%22&st=cse. Accessed March 27, 2011.

Halpern, Benjamin S., et al. "A Global Map of Human Impact on Marine Ecosystems." *Science* 319, no. 5,865 (15 February 2008): 948–952. Available online. URL: http://www.edf.org/documents/7646_oceansmap_science_2_15_08.pdf. Accessed on March 27, 2011.

Laurance, William F. "When Bigger Is Better: The Need for Amazonian Mega-Reserves." *TRENDS in Ecology and Evolution* 20, no. 12 (December 2005): 645–648.

Magro, Teresa Christina, Alan Watson, and Paula Bernasconi. "Identifying Threats, Values, and Attributes in Brazilian Wilderness Areas." USDA Forest Service Proceedings RMRS-P-49 (2007): 319–322. Available online. URL: http://www.fs.fed.us/rm/pubs/rmrs_p049/rmrs_p049_319_322.pdf. Accessed March 27, 2011.

Olsen, Erik. "Protected Reef Offers Model for Conservation." *New York Times* (26 April 2010). Available online. URL: http://www.nytimes.com/2010/04/27/science/earth/27reef.html?_r=1&scp=1&sq=%22protected+reef%22&st=nyt. Accessed March 27, 2011.

Pala, Christopher. "A Struggle to Preserve a Hawaiian Archipelago and Its Varied Wildlife." *New York Times* (19 December 2006). Available online. URL: http://www.nytimes.com/2006/12/19/science/earth/19hawa.html. Accessed March 27, 2011.

———. "No-Fishing Zones in Tropics Yield Fast Payoffs for Reefs." *New York Times* (17 April 2007). Available online. URL: http://www.nytimes.com/2007/04/17/science/earth/17fish.html?_r=1&scp=13&sq=%22no+take%22&st=nyt. Accessed March 27, 2011.

Palmer, Todd M., et al. "Breakdown of an Ant-Plant Mutualism Follows the Loss of Large Herbivores from an African Savanna." *Science* 319, no. 5,860 (11 January 2008): 192–195.

Pew Environmental Group. "World's Marine Scientists Call for Large-Scale 'National Parks at Sea.'" (8 June 2010). Available online. URL: http://globaloceanlegacy.org/newsroom/release_8june2010.html. Accessed March 27, 2011.

Peres, Carlos A. "Why We Need Megareserves in Amazonia." *Conservation Biology* 19, no. 3 (June 2005): 728–733. Available online. URL: http://www.uea.ac.uk/~e436/Peres_ConBio2005.pdf. Accessed March 27, 2011.

Revkin, Andrew C. "Human Shadows on the Seas." *New York Times* (26 February 2008) Available online. URL: http://www.nytimes.com/2008/02/26/science/earth/26coas.html. Accessed March 27, 2011.

Sandin, Stuart A., et al. "Baselines and Degradation of Coral Reefs in the Northern Line Islands." *PLoS ONE* 3, no. 2 (27 February 2008): e1548. Available online. URL: http://www.plosone.org/article/info%3Adoi%2F10.1371%2Fjournal.pone.0001548. Accessed March 27, 2011.

Siebert, Charles. "An Elephant Crack-up?" *New York Times* (8 October 2006). Available online. URL: http://www.nytimes.com/2006/10/08/magazine/08elephant.html?scp=1&sq=elephants and acacia&st=cse. Accessed March 27, 2011.

Soulè, Michael E., and John Terborgh, eds. *Continental Conservation: Scientific Foundations of Regional Reserve Networks.* Washington, D.C.: Island Press, 1999.

Wilderness Foundation South Africa. "Baviaanskloof Mega Reserve Project." (August 2008). Available online. URL: http://www.wildernessfoundation.org.za/project/content.asp?PageID=103. Accessed March 27, 2011.

Web Resources

For the latest information on the environmental topics considered in this volume, readers can consult the Web sites of the following government agencies, associations, nonprofit organizations, and professional journals. All provide searchable text online and may also provide interactive features, video clips, podcasts, and links to other relevant sources.

American Association for the Advancement of Science (AAAS). URL: http://www.aaas.org/. News related to a broad range of scientific topics and careers. Accessed March 27, 2011.

American Journal of Bioethics. URL: http://www.bioethics.net/. Free access to abstracts and some full text articles, online discussions, and links to relevant news articles on a range of topics in bioethics, including animal treatment issues. Accessed March 27, 2011.

Conservation International. URL: http://www.conservation.org. Information on global initiatives to protect biodiversity and terrestrial hotspots. Accessed March 27, 2011.

Convention on Biological Diversity (CBD). URL: http://www.cbd.int/. Links to news and information about the international treaty. Accessed March 27, 2011.

Foundation for Deep Ecology. URL: http://www.deepecology.org/. Information on sustainable development and the deep ecology movement. Accessed March 27, 2011.

Humane Society of the United States (HSUS). URL: http://www.hsus.org/. Information about the treatment and protection of animals in the United States. Accessed March 27, 2011.

International Union for Conservation of Nature (IUCN). URL: www.IUCN.org. Access to information on a wide range of conservation issues, including the IUCN Red List of Threatened Species. Accessed March 27, 2011.

Natural Resources Defense Council. URL: http://www.nrdc.org/. Articles and action ideas on a variety of environmental topics. Accessed March 27, 2011.

Proceedings of the National Academy of Sciences (PNAS). URL: http://www.pnas.org/. Open access to articles on a broad range of scientific topics. Accessed March 27, 2011.

Public Library of Science (PLoS). URL: http://www.plos.org/. Open access to articles on a variety of topics in the biological and environmental sciences. Accessed March 27, 2011.

United Nations Collaborative Programme on Reducing Emissions from Deforestation and Forest Degradation in Developing Countries (UN-REDD). URL: http://www.un-redd.org/. Information on international economic and technical assistance for developing countries following low-carbon paths to sustainable development. Accessed on March 27, 2011.

United Nations Environment Programme (UNEP). URL: www.unep.org. Resources on a wide array of global environmental issues and initiatives. Accessed March 27, 2011.

U.S. Department of Agriculture (USDA). URL: www.usda.gov. Information on domestic agricultural practices, including organic farming and sustainable agriculture initiatives. Accessed March 27, 2011.

U.S. Department of the Interior. URL: www.doi.gov. Links to information on domestic wildlife issues through the U.S. Fish & Wildlife Service, the U.S. Geological Survey, and the National Park Service. Accessed March 27, 2011.

U.S. Environmental Protection Agency (EPA). URL: www.epa.gov. Resources on a range of pollution issues and cleanup initiatives. Accessed March 27, 2011.

World Bank. URL: www.worldbank.org. Information on sustainable development and environmental projects in developing countries. Accessed March 27, 2011.

World Wildlife Fund (WWF). URL: http://www.wwf.org/. Information on threatened and endangered species, their habitats, and opportunities for environmental action. Accessed March 27, 2011.

Free Online Print and Radio Media

Several of the print and radio news sources referenced in this volume are available free to online users. The following Web sites contain bonus audiovisual content, such as video clips, slide shows, interactive graphics, and podcasts.

National Public Radio (NPR). http://www.npr.org/. Accessed March 27, 2011.
Newsweek. URL: http://www.newsweek.com/. Accessed March 27, 2011.
New York Times. URL: http://www.nytimes.com/. Accessed March 27, 2011.
San Francisco Chronicle. URL: http://www.sfgate.com/. Accessed March 27, 2011.
ScienceDaily. URL: http://www.sciencedaily.com/. Accessed March 27, 2011.
Slate. URL: http://www.slate.com/. Accessed March 27, 2011.
TIME. URL: http://www.time.com/time/. Accessed March 27, 2011.
Washington Post. URL: http://www.washingtonpost.com/. Accessed March 27, 2011.

INDEX

Note: *Italic* page numbers indicate illustrations. Page numbers followed by t denote tables, charts, or graphs; page numbers followed by m denote maps; page numbers followed by c denote chronology entries.

A

acacia trees 11, 101, 106–110, *107*
acetaminophen 88
activism 64
Africa 106–110, 119c
Agriculture, U.S. Department of 45, 88
Alabama 93
algae 95–96
Alley, Richard 98–99
Amazon rain forest 117c–118c
    biodiversity in 7
    deforestation 34m, 35t, 37t
    economic incentives for preservation 36–38
    extent of devastation 34m, 34–36
    habitat destruction 30–40, *32*, *33*, 35t, 37t
    mega-reserve proposals 112–113
    restoration 39–40
ANDi (genetically modified rhesus monkey) *98*
animal-dispersed trees 39

Animal Liberation (Singer) 14–15, 116c
animal rights 13–15
Ankole cattle *11*
Antarctic minke whale *71*
anthropocentric arguments for conservation of biodiversity 12–13
apples 3
Arabian oryx 78
archipelago 1, 4, 118c, 119c
Arctic Circle 4
Arctic sea-ice
    future projections 26t
    polar bear habitat 25–27
    summer minimums 24m, 118c
Aristotle 12
Asian carp 10, 85, 88–90, *89*, 119c
Australia 72, 74, 104

B

bajii 8
bald eagles 47–49, *51*, 115c, 118c
Baltzer, Mike 76
Bartlit, Fred, Jr. 53–54
Baviaanskloof Mega-reserve (South Africa) 112
Belize 104
Bentham, Jeremy 15, 115c
benzene 65
bighead carp. *See* Asian carp
biocentrism 15–16, 20–21
biodiversity hot spots. *See* hot spots

biomagnification 48, *50*
biomass 7
biomes 112
biotechnology 18–20
birds 86–87
black-footed ferrets 77
blacktip reef sharks *102*
bluefin tuna *57*
Boiga irregularis 87
BP oil spill (2010) xiv–xv, *52*, 52–58, *53*, 120*c*, 121*c*
  extent of destruction xv, 56–58, 64–65, 67
  marine animal deaths 55*m*
  underwater effects *57*
Bradshaw, Gay 109
Brazil 30–32, 118*c*
BRCA1/BRCA2 gene 20
Brown, James 91
brown pelicans *57*
brown tree snakes 10, 85–88, *87*
buffalo 10
Burger, Warren 18, 116*c*
Burnett, Joe 49
Bush, George W., and administration 27, 80, 118*c*, 120*c*

C

California condors 47, 49, 77, 120*c*
Camilli, Richard 58, 65
Canada thistle 96
cap-and-trade plans 38
captive breeding 77–79
carbon dioxide (CO_2) 35–36, 38, 119*c*
carbon sequestration 37
carbon trading 38, 119*c*
Carcharhinus melanopterus *102*
carp 10, 85, 88–90, *89*, 119*c*
Carson, Rachel 9–10, 42–47, *46*, 49, 50, 115*c*
The Case for Animal Rights (Regan) 14
cats 93

Center for Biological Diversity 27
cetaceans 72–74
China 76, 77, 118*c*, 119*c*
chlorinated hydrocarbons 47
Chukchi Sea 22, 118*c*
CI. *See* Conservation International (CI)
Clean Air Act 30
climate change. *See* global warming
coffee 40
cogongrass 93–94, *94*
condors 47, 49, 77, 120*c*
Conklin, Eric 95, 96
conservation hot spots 28–31, 29*m*
Conservation International (CI) 28, 30, 76, 117*c*
conservation of biodiversity 1–21
  animal rights arguments for 13–15
  anthropocentric arguments for 12–13
  biocentric/deep ecology arguments for 15–16, 20–21
  causes of extinction 5–11
  ethical, aesthetic, and economic reasons for 12–17, 20–21
  Global Seed Vault 1–5, *2*, *3*, 118*c*
Continental Conservation (Soulè and Terborgh) 112
Copenhagen Accord 30, 38
coral reefs
  and ecosystem disruption 11, 101
  restoration of 95–96
  and sharks *102*, 102–106
coral trout 104
Crabtree, Robert 96
Cretaceous–Tertiary (K-T) extinction event 6
Crichton, Michael 18–19
critical habitat 27, 118*c*, 119*c*
crop diversity 3–4

Index

D

DDT 43, 44
  bald eagles and 47–49, 118c
  banning of 47–52, 116c, 118c
  biomagnification 50
  California condor and 47, 49, 120c
  chemical structure/breakdown 48
  ecological stress from 9–10
  Silent Spring and 42–47
deep ecology 15–16, 20–21, 116c
Deepwater Horizon explosion and oil spill (2010). *See* BP oil spill
deforestation 34m, 35t, 36–38, 37t
detergents 61
DiCaprio, Leonardo 79
dichlorodiphenyltrichloroethane. *See* DDT
Dick, Christopher 40
Dinerstein, Eric 79–80
disruption 9
DNA 18
dolphins 57, 74
doomsday vault 1–5, 2, 3, 118c
Drengson, Alan 20–21

E

ecosystem disruption 10–11, 101–110
  acacia trees 106–110
  coral reefs 102, 102–106
  elephant culture 108–109
  as human cause of extinctions 10–11
mega-reserves as cure for 110–113
ecosystems, invasive species' positive influence on 90–93
elephants 11, 107, 108–109
Elsa (lioness) 79
emmer wheat 5
endangered species
  bald eagles 47–49, 115c, 118c
  California condors 47, 49, 120c
  gray wolves 80, 80–83, 83m, 119c–121c
  IUCN Red List 17t, 49, 69
  polar bears 22–30, 23, 118c, 119c
  tigers 76, 78–80, 119c
  whales 70–75, 71, 73t, 75t
Endangered Species Act 27, 82, 116c, 118c
Endeavor (research ship) 58
endemic species 28, 117c
Environmental Protection Agency (EPA) 45, 116c
extinction(s) 6t. *See also* endangered species; threatened species
  captive breeding to prevent 77
  causes of 5–11
  Endangered Species Act and 116c
  human population growth and 9
  invasive species and 86, 91–93
mega-reserves to prevent 112, 117c
extirpation 10, 16
Exxon Valdez oil spill xv, 58, 66, 67, 117c

F

factory ships 70
FAO. *See* Food and Agriculture Organization (FAO)
fertilizers 60
fish 79
Fish and Wildlife Service, U.S. 81
Food and Agriculture Organization (FAO) 34
forests 92–93, 115c
Forest Service, U.S. 12
Fowler, Cary 2, 4
fragmentation 9
Friedman, Thomas 35–36

G

Gaines, Steven 91
genes, patenting of 18–20, 116c
genetic engineering 18–20, 98, 116c
giant pandas 77
Global Canopy Program 38
global climate change. *See* global warming
Global Crop Diversity Trust 2, 5
Global Ocean Legacy 106
Global Seed Vault. *See* Svalbard globale frøhvelv
global warming
    Arctic sea-ice summer minimums 24m, 118c
    change in species' ranges 99, 117c
    habitat loss 9
    invasive species 94, 96, 98–99
    polar bear habitat destruction 22–30, 118c
Glover's Reef 104, 110
Gore, Al 43–44
Granholm, Jennifer 90
grass carp. *See* Asian carp
gray wolves 10, 80, 80–83, 83m, 119c–121c
Great Barrier Reef Marine Park (Australia) 104, 110
Great Lakes 86, 88–90, 119c
greenhouse gas 27, 96
Greenpeace 27
Grimsley, Sean 54
grizzly bear 13
Guam 78, 86–88
Guam rail 77–78
Gulf of Mexico oil spill (2010). *See* BP oil spill
Guyana 36, 38, 119c

H

habitat destruction 22–41
    Amazon rain forest 30–40, 32, 33, 34m, 35t, 37t
    polar bears and global warming 22–30
habitat loss 8–9
Halliburton 53–54, 56
Harvard University 116c
Hawaiian Islands 95–96
Hayward, Tony 64
HEC. *See* human-elephant conflict
Henderson, Henry 89
herbivores 106
hippocampus 108
Holder, Eric H., Jr. 56
hot spots (biodiversity/conservation) 7, 28–31, 29m, 33, 116c–117c, 117c
household chemicals 59–60
Howe, Henry 39
human-elephant conflict (HEC) 109
human impact on oceans 105m
hydrocarbons 47, 65
Hynes, H. Patricia 44–45

I

Idaho 81, 119c–120c
indicator species 7
indigenous populations 4
Intergovernmental Panel on Climate Change (IPCC) 25
Interior, U.S. Department of the 27, 67, 80, 120c, 121c
International Union for Conservation of Nature (IUCN) Red List of threatened species 17t, 49, 69
International Whaling Commission (IWC) 70, 72, 74, 116c
introduced species 10. *See also* invasive species
    invasive species 10, 85–100, 119c
    Asian carp 88–90, 89
    brown tree snake 86–88, 87
    climate change and 96, 98–99
    distribution of marine species 97m

Index

positive influence on ecosystems 90–93
and restoration of coral reefs 95–96
and restoration of forests 92–93
IPCC. *See* Intergovernmental Panel on Climate Change (IPCC)
IUCN. *See* International Union for Conservation of Nature (IUCN)
IWC. *See* International Whaling Commission (IWC)

J

Japan 72, 74, 120c
Jha, Shalene 39–40
Joye, Samantha 65
Justice Department, U.S. 20, 56

K

Kareiva, Peter 31
Keiko (orca whale) 78
keystone species
 defined 7
 and ecosystem disruption 10–11
gray wolves 81–82
Kiss, Agnes 30–31
K-T (Cretaceous–Tertiary) extinction event 6

L

land ethic 15–16, 114, 115c
Landry, Mary E. xiv
landscaping 61
Laurance, William 112–113
Leopold, Aldo 15–16, 101–102, 114, 115c
lions 79
livestock 34
Lubchenko, Jane 65, 67

M

MacDonald, Ian 65
Macondo exploratory well 52
Macquarie Island 93
macroecology 91
maize 3
Marine Mammal Protection Act 27
Marvier, Michelle 31
mass extinction events 5–7
Matthiessen, Peter 45–46
mega-reserves 110–113, 111m
Mendes, Francisco Alves (Chico) 38
Michigan, Lake 89, 90, 119c
Mills, Judy 76
Minerals Management Service 67–68, 120c
minke whale 71
Mitchell, Andrew 38
Mittermeier, Russell 31
Molloy, Donald W. 81, 82
monocultures 39
Montana 81, 82, 119c–120c
Montgomery, Tony 95
Montrose Chemical Corporation 49
moratorium, whaling 70, 116c
Morishita, Joji 72
Moura, Vitalmiro Bastos de 31
Muir, John 15
mutualism 106–110
Myers, Norman 28, 30, 116c–117c
mysticetes 74

N

Naess, Arne 20, 116c
Nany River 30–40, 32
National Ocean Council 68, 120c
National Oceanic and Atmospheric Administration (NOAA) 25, 58, 64, 65
national parks 110–111, 119c
National Science Foundation (NSF) 39
Natural Resources Defense Council (NRDC) 27

Nature Conservancy 95
New York City 92
New Zealand 91, 93
niche crops 4
niches 41, 91, 93, 96
NOAA. *See* National Oceanic and Atmospheric Administration (NOAA)
Norway 72. *See also* Svalbard globale frøhvelv (Global Seed Vault)
Norway maple 92
no-take reserves 104
NRDC. *See* Natural Resources Defense Council (NRDC)
NSF. *See* National Science Foundation (NSF)

O

Obama, Barack, and administration
  BP oil spill investigation 53
  gray wolf policy 80, 120c
  Minerals Management Service policies 67–68, 120c
  offshore drilling policies 67, 120c–121c
  polar bear habitat designation 27, 119c
oceans, human impact on 105m, 118c
odontocetes 74
offshore drilling 121c. *See also* BP oil spill
oil spills xiv–xv, 61, 66
OncoMouse 18, 19, 116c
orca whales 78
organisms, patenting of 18–20, 116c
organochlorides 47
oryx 78
otters 68
overexploitation 10

P

Palmer, Todd 106
Palos Verdes Shelf 49
pandas 77, 118c
patenting of genes/organisms 18–20, 116c
Patent Office, U.S. 18, 20, 116c
peas 3
Pecot, Stephen 94
Peres, Carlos 112–113
Permian-Triassic mass extinction 5
pesticides 115c. *See also* DDT
phytoplankton 65
Pikitch, Ellen 104
Pinchot, Gifford 12–13, 115c
plankton 57, 89–90
PNAS. *See Proceedings of the National Academy of Sciences (PNAS)*
poaching 10
polar bears 22–30, 23, 118c, 119c
Politics (Aristotle) 12
pollution 9–10
population declines in vertebrate species 6t
  predators
  and biomagnification 48
  brown tree snakes 86
  and coral reef health 103
  gray wolves 81–82
invasive species 91
preservation, economic incentives for 36–38
Prince William Sound, Alaska 66, 117c
Pringle, Robert 107, 110
Proceedings of the National Academy of Sciences (PNAS) 36
python 78

R

rabbits 93
rain barrels 62

Index

rain forests. *See also* Amazon rain forest
  biodiversity in 7
  restoration 39–40
rain gardens 61–62
recycling 61, 62
REDD (Reduced Emissions from Deforestation and Degradation) 119c. *See also* UN-REDD
Red List of threatened species 17t, 49, 69
reforestation 39–40
Regan, Tom 14
Ricciardi, Anthony 92–94
rights, legal/philosophical definition 14
Rogers, Haldre 86, 88
Roosevelt, Theodore 13
Root, Terry 99
Rowan, Andrew 14

S

Saipan 87–88
Sala, Enric 103, 104, 106
Salazar, Ken 27, 30, 67, 68, 81
Salvat, Bernard 106
A Sand County Almanac (Leopold) 15, 115c
Sax, Dov 90–91
Schlickeisen, Rodger 82
sea-ice. *See* Arctic sea-ice
sea lions 49–50
sea otters 68
sea turtles 57
seed banks 4
seed dispersal 87–88
seed vault. *See* Svalbard globale frøhvelv (Global Seed Vault)
sentient creatures 13
Sessions, George 20–21
shade coffee farms 40
sharks 11, *102*, 102–106
Shorett, Peter 19

shrimp 57
Sierra Club 15
Silent Spring (Carson) 10, 42–47, 115c
Silva, Tarcísio Feitosa da 32
silver carp. *See* Asian carp
Simpson, Mike 81
Singer, Peter 14–15, 116c
Skovmand, Bent 19–20
Soulè, Michael 112
South Africa 109, 112
Southern Ocean 74, 120c
species extinctions. *See* extinctions
speciesism 15
species loss 69–84
sperm whales 57
Stang, Dorothy 30–32, 117c–118c
steelhead trout 79
stem-boring beetles 107
storm drains 61
Stuart, Bruce 92
suffering, animal rights and 13–15, 115c
Super Sucker 95, 96
Supreme Court, U.S. 18, 90, 116c
sustainable development 36–38
Svalbard globale frøhvelv (Global Seed Vault) 1–5, *2*, *3*, 19, 118c

T

Terborgh, John 112
terrestrial species 7
Tester, Jon 81
Thoreau, Henry David 15
threatened species, IUCN Red List 17t, 49, 69
tigers 10, 76, 78–80, 119c
  toxic contamination and cleanup 42–68
  BP oil spill 52–58, 64–65, 67
  DDT ban 47–52
  Silent Spring and DDT 42–47

The Training of a Forester (Pinchot) 13, 115c
Transocean 53, 54, 56
trash disposal 62
tree snakes 86–88, *87*
trout 79
Tylenol 88

U

umbrella species 7
understory species 40
United Nations Conference on Climate Change 36, 119c
United Nations Food and Agriculture Organization (FAO) 34
University of Hawaii 95
UN-REDD (The United Nations Collaborative Programme on Reducing Emissions from Deforestation and Forest Degradation in Developing Countries) 38, 41, 119c
utilitarianism 15

V

vascular plants 28, 36
Veracruz, Mexico 39
vertebrate species, population declines in 6t

W

Walden (Thoreau) 15
water conservation 62, 64
water quality, improved 59–64, 63t
weeds 99
wetlands 31
whales and whaling 70–75, *71*, 73t, 75t, 115c, 116c, 120c
Whitehead, Hal 72–73
White-Stevens, Robert 46
Woods Hole Oceanographic Institution (WHOI) 58, 65
World Wildlife Fund (WWF) 23, 25, 76, 79
Wyoming 80–81

X

Xiang Xiang (giant panda) 77, 118c

Y

Yangtze River dolphin *8*
Yellowstone National Park 82, 96, 98
Yi, Heng 77
Yohe, Gary 99
York, Geoff 23, 25
Yushin Maru 71

Z

Zavaleta, Erika 93